THE
NINETY-MINUTE
HOUR

ALSO BY JAY CONRAD LEVINSON

Guerrilla Marketing Attack
Guerrilla Marketing
Quit Your Job
An Earthling's Guide to Satellite TV
Small Business Savvy
555 Ways to Earn Extra Money
Earning Money Without a Job

JAY CONRAD LEVINSON

THE NINETY-MINUTE HOUR

E. P. DUTTON NEW YORK

Published in the United States by E. P. Dutton, a division of Penguin Books
USA Inc., 2 Park Avenue, New York, N.Y. 10016.

Published simultaneously in Canada by
Fitzhenry and Whiteside, Limited, Toronto.

Library of Congress Cataloging-in-Publication Data
Levinson, Jay Conrad.
The ninety-minute hour/Jay Conrad Levinson. — 1st ed.
p. cm.
ISBN 0-525-24851-X
1. Time management. I. Title.
HD69.T54L48 1990
650.1'2—dc20   89-37806
                                            CIP

Designed by Margo D. Barooshian

3 5 7 9 10 8 6 4 2

This is for Patsy,
because Patsy is the person
with whom I most love to share
my time.

Contents

CONTENTS

PART ONE

THE IDEA

1

The Ninety-Minute Hour
in Action

Tom Dennison is the president of a profitable and growing company. It happens to be a publishing company, but
it could be any company.

Tom lives with his wife, Kathleen, and two kids, Nan,
fourteen, and Grant, ten, in a restored 1820 farmhouse on
185 acres of land. The property is secluded, lush, and beck-
oning. Two year-round steams flow through it. And it's
only fifty-five minutes away from the city in which his
office is located.

You can't blame Tom for wanting to spend as much
time as possible in his beautiful country setting. So you
probably won't be too surprised that he does—three days

a week. His workweek begins the moment he leaves home at 8:00 Monday morning. It ends with his arrival at home, around 6:00 Thursday evening.

That seems like a lot of free time for the president of a growing company. It is a lot, that's true. And you don't even know about Tom's two monthlong vacations each year. But that's not important now. What is important is the way that Tom considers time.

Tom considers time to be one of life's most precious assets—ranking right up there with health and love. Time is far more valuable to Tom than money or fame or travel. To Tom, time is really another word for "life." As far as he's concerned, time for him will end when his life ends— just as it began when his life started. Sure, he deeply loves Kathleen, Nan, and Grant. They're the most important people in his life. But he recognizes that his life, and theirs, is really composed of time—large and small pieces of time that have been artificially divided into segments called minutes and hours.

So Tom knocks himself out to waste little time and to spend what time is his as valuably as he possibly can. In pursuit of this goal, Tom gives serious and conscious thought to the expenditure of time. He wants to get the most out of his working time, his playing time, and his relaxation time.

Tom fully realizes that most people don't share his views about time. It's not that they're against them. It's more that they haven't gotten around to recognizing the crucial nature and limited supply of time available to them. Some of these people organize their time. That's a step in the right direction, but it's only a step. It's not the whole distance. Tom tries to travel that whole distance.

Here's how he goes about it: On Monday morning, Tom dictates two memos into his microcassette player-

4

recorder while driving to the office. As soon as he gets there, he slips a soft jazz cassette into the stereo unit in his office. He asks Jerry Daniels, his production vice president, to attend the Boone meeting for him and to negotiate the Baker printing deal afterward. Baker is coming to the office to finalize the terms of the deal.

Tom asks Sandra Masters, his secretary, to answer three letters, following the format he's established. She'll do a perfect job; she always does. Tom trained her well. Truth is, she answers about 90 percent of his mail.

The music in Tom's office soothes as it implants in Tom's mind subliminal suggestions on goal setting. It is fitting background music as Tom makes a proposal to a major chain of bookstores, using his computer, telephone, and modem.

Next on Tom's agenda is a meeting with a videotape production company that wants to talk merger with Tom's publishing company. Because he wants to view their operations in person, the meeting is arranged to take place in their office, a thirty-minute drive away. While driving there, Tom speed-listens to an entire sixty-minute cassette on creative problem solving. After the meeting Tom drives back to his office, using his car phone to return three calls and make three more on the way.

After an unhurried lunch, Tom speed-reads his business mail, three business magazines, and two newsletters. His computer prints out an acceptance of his proposal by the bookstore chain. This means Tom can fly to Japan the next day to meet with the president of a large Japanese publishing firm.

He schedules the tasks he'll accomplish while flying both ways and makes sure he has his Japanese language instructional tapes for the drive home later—also for part of the plane trip.

After dinner and plenty of satisfying time with the kids, Tom and his wife watch three hours of high-quality TV. It takes them only two hours, courtesy of their VCR, satellite dish, and merciless zapping of the commercials—completely eliminating them by pushing the fast-forward button on the remote control switch whenever they appear during program breaks or between shows.

The final forty-five minutes of television are spent watching a comedy tape by one of Tom and Kathleen's favorite performers. While the tape is filling the room with outrageous humor, it's also putting forth subliminal messages on negotiating tactics.

Tom isn't the president of a successful company by accident. He enjoys the benefit of having more time than ordinary people. He has become a master in the art of time extension—doing one thing while doing another, delegating tasks, and speeding up the time it takes to do many activities. As a result, Tom's hours have ninety minutes in them.

Tom Dennison's ninety-minute hours are one of the prime reasons that his company is so successful. Tom's extra time is also the secret of his brief workweek. It is the key he has used to master so many of the skills necessary to run a prospering company. It's the way he can afford to take so many luxuriously long vacations. Tom's extra time also gives him extra enjoyment at home with his family. Tom's hours have ninety minutes of time and of living in them.

There is little question that Tom Dennison is a very wealthy man. But I'm not talking about his bank account or his company's cash flow. I'm not referring to his CDs, his stocks, his tax-free municipal bonds, or even his real-estate holdings. Instead, I'm focusing on Tom's time. Not only does he have more of it than most people, but he also has a system to create even greater quantities of time whenever he needs to. Tom considers time his most precious possession, and he's loaded with it. Filthy rich in time, that's Tom.

That sure doesn't describe Stan Danton, though. Stan also happens to be the president of a publishing company. Stan lives with his wife, Marcia, and two kids, Cindi, sixteen, and Mark, eleven, in a restored farmhouse on 160 acres of land. Like Tom's house, Stan's is only fifty-five minutes away from the city in which he maintains his company office.

Stan is nuts about his home and its pastoral surroundings. He'd love to spend more time there, but it's just plain impossible to do. After all, running a company is a big drain on his time. So you probably won't be too surprised that he spends at least five full days a week at the office.

Stan's workweek begins the moment he arrives at his office at 9:00 Monday morning. It ends when he leaves the office around 6:30 Friday evening. But that's not really the end. Stan puts in about ten more hours working on business matters at home each week. It's pretty apparent that Stan doesn't have much free time. But after all, everyone knows that running a company takes more than a nine-to-five five-day week.

Statistics tell us that top executives work the equivalent of fifteen months a year. Stan crowds his work into twelve months. Tom, with his two monthly vacations, accomplishes his in ten months. And don't forget—Tom's months are filled with four-day weeks, not the five-day weeks that govern Stan's life.

The average top executive also spends ten and a half hours weekly working at home. That's more than thirteen extra forty-hour weeks each year. More proof that time spent working is a requisite for high achievement: the average top executive puts in a total of fifty-seven hours weekly on the job, counting work in the office and at home.

Stan Danton, working forty-seven and a half hours at the office, plus about ten more hours at home, is devoting near the average amount of time per week to his work. But countless people feel compelled to put in far more than the average. I'm talking about six- and even seven-day weeks. I'm talking about arriving at work at 8:00 A.M. and leaving about 8:00 P.M. or later. I'm talking about a whole lot of people, too.

The people who put in those brutal hours feel they have no choice but to be work bound. That's certainly the way Stan feels. He doesn't necessarily like it, but he believes that extra hours are the dues you must pay when you reach the top.

THE MYTH OF THE SIXTY-MINUTE HOUR

Of course, Stan also believes that one hour contains sixty minutes and no more. He's never heard of time extension. His weekends are always two-day weekends, sometimes one-day and no-day weekends, hardly ever three-day weekends, except for the occasional national holiday. And he's never had a vacation that lasted a month, let alone two extended vacations in one year. Stan doesn't see how he'll ever get a lengthy vacation like that before retirement.

The truth is, free time and long vacations have never entered Stan's mind. But three-day weekends and luxurious vacations aren't important now. What is important is the way that Stan considers time. To be blunt about the whole thing, he really doesn't consider it much. Oh, he organizes it. He's read a book about time management. And he even keeps a daily calendar of sorts. But Stan, like most business executives—like the majority of people—doesn't really realize the precious and elusive nature of time. Sadder still, he is unaware of his ability to control time, to extend it, to put it to work *for* him rather than *against* him. If Stan were ever asked to list his most valuable assets, you could be sure that time would find no place on the list.

To Stan, time is the enemy, not the friend. It is something he works against rather than with. He does appreciate many of the finer things in life, but he doesn't see the connection of time to life. As far as he's concerned, time has always been and will always be the same. Stan Danton, an educated and intelligent man, never has given much thought to time. So he is not oriented to saving it or using it in ways other than the ways his ancestors used it—working hard and long hours.

Unlike Tom Dennison, Stan rarely if ever gives serious and conscious thought to the expenditure of time. Somewhere in the back of his mind, he may be aware of the crucial nature of time, also of its limited supply in his life. But what the heck, he works hard and he figures that's enough. After all, if you put in an average of fifty-seven hours working each week, wouldn't you figure that was enough?

DAILY CALENDARS ARE NOT ENOUGH

When Stan started keeping a daily calendar, he figured he was mastering time. There's no doubt that he was heading in the right direction. But he was a long distance from going all the way. Unfortunately, Stan thought that was as far as he had to go when it came to taking control of time.

But here's how far a calendar really takes him: On the way to work Monday morning, Stan listens to his favorite talk-radio show on his car stereo. As soon as he arrives at the office, he notices on his calendar that he's got an important production meeting to attend. As if that weren't enough, he's also got to negotiate a major printing deal, the Addison deal, that afternoon. And Addison himself is coming to the office.

Stan asks Martha Winters, his secretary, to come into his office so he can dictate three letters to her. Answering his mail is one of his least enjoyable chores, but after all, someone's got to do it.

One of the pieces of mail he attends to is a brochure about a seminar on goal setting. Stan signs up for the Saturday session. Sure, it's going to mean one more day away from his home and family, but goal setting is an important art, especially for a company president.

He asks Martha to take all calls and fend off all visitors while he makes the notes for a proposal to a major chain of bookstores, using his trusty ballpoint pen and yellow, lined legal pad. Stan never did learn to type. Later, he'll dictate the proposal to Martha, who will type it and then mail it.

Next in his day is the production meeting. It's a long one because of all the minute details of production. Still, if he didn't attend it, who could?

Because of the meeting, business correspondence, and the impending negotiating appointment with Addison, who will be arriving at 2:00 in the afternoon, Stan cancels his plan to meet with the videotape production company that wants to talk merger with his publishing company. He decides to meet with them later in the week. Friday afternoon is the time set for a meeting at their office. Sure, rush-hour traffic will be peaking, but business is business.

After a quick lunch, it's time for the negotiating meeting. It's going to be a long and tough session, because Addison is a crafty negotiator.

There is so much to do as president of a company that there never seems to be enough time to give proper attention to all the normal problems of business. That's probably why Stan decides to sign up for a two-day course on creative problem solving. Yes, it will mean two more days away from his family, two more days away from his home, but how else is a person to learn creative problem solving?

Following his meeting with Addison, Stan spends an hour on the phone. Three phone calls are calls he must return; three more are calls he must make.

Then comes the time that Stan, as a conscientious company president, must wade through his business mail. There's so much that he scarcely has time to scan a couple of business magazines, and he has no time for the newsletters that have arrived.

It would benefit his company if Stan could take the time to enter into a joint publishing venture with a Japanese publishing firm that has sent out feelers, but where's he going to find the time to go to Japan? And what about the language barrier? Sure, other people have tackled those dilemmas, but when can Stan take the time to work out solutions?

Eventually, Stan arrives home. On the way he relieved some of the day's tensions listening to music in his car. The kids having finished dinner, Stan dines with his wife. After dinner he apologizes to Cindi for not being able to see her in the school play and also asks Mark's forgiveness for not having the time to help with his science project.

Stan still has about an hour and a half of work to do before he can unwind watching a few hours of TV with his wife, viewing whatever fare is served up that night— some pretty good, some pretty mediocre. At least it's a way of turning off his mind. Stan falls asleep after midnight while reading a book on negotiating tactics.

Unfortunately, Stan never got around to many of the things he wished to accomplish that day. No wonder. His hours have but sixty minutes in them. His time is as limited as the next guy's—unless the next guy happens to be Tom Dennison.

THE LIMITATIONS OF SIXTY-MINUTE HOURS

Stan Danton's sixty-minute hours are keeping his company from the growth it could be enjoying. Those same sixty-minute hours are the reason his workweek is so long. And because he feels that he must master new skills as the president of a company, he spends even more sixty-minute hours in his quest for knowledge. No wonder he has no time for lengthy vacations. No wonder people consider him a workaholic. No wonder Stan is somewhat of a stranger to his own kids.

Some people might consider Stan a fairly wealthy man. If you consider his home, his company, and his savings account, you might be impressed with his wealth. But if you looked deeper and saw the way Stan is living his life, you'd realize that his work and lack of understanding of time have impoverished the man. He only has the trappings of wealth. The reality is that Stan has been living on the edge of time bankruptcy, never having a surplus of time, constantly enslaved by his lack of it.

What Stan needs so desperately is something that no amount of money can buy. He needs more time. He needs more minutes in each hour. He needs a system to give him constant access to those minutes. He needs ninety-minute hours.

Slowly, very slowly, it is dawning on Stan that he lacks time. Slowly, he is realizing that he can get it. And it's not even all that difficult.

The first thing Stan must comprehend is that new findings in technology and psychology have given human-kind a new degree of control over time. He must learn that it is possible to do a myriad of things faster than ever before, to put ninety minutes into his hours. Then, he must determine how to put the maximum number of ninety-minute hours into his life.

TOOLS FOR THE NINETY-MINUTE HOUR

To achieve the goal of gaining more time, at least nine tools are now available. One is old and obvious, but beyond the capability of many of the people who need it most. Some are technical tools. Some are behavioral tools. Some are psychological breakthroughs. Some are commonplace items that, when used correctly, unlock the door to the treasure of extra time. There will be increasing numbers of tools every year from now on, as we realize that we have the ability to harness time.

In upcoming chapters I will describe these tools in detail, and I will show you how to make them your tools, your methods of enjoying ninety-minute hours. But now let's take a fleeting glimpse at them so that you can start priming yourself for a new, accomplishment-filled life—a life that can benefit you, your family, and your business.

DELEGATION: THE GREAT UNTAPPED TIMESAVER

The most ancient of the tools available to you is the fine art of delegating. Don't underestimate the timesaving power of this talent—the surprisingly rare ability to enable others to handle your work with the same quality you'd put into it. People know about delegating. It's just that they're too busy to do it.

Some people have egos too large to allow them to delegate. They figure that if they don't do it themselves, it won't get done right, or at all. Or they may assume that the mere act of delegating will make them appear extraneous to the operation. Or they may enjoy doing the task, so they do it even though there are far more important things for them to do. That's a shame, because while they're getting their jollies, their company is being deprived of their other abilities.

Still other people fail to delegate simply because they were never taught how. Once, when I was a junior executive in a large company, my secretary asked if she could have a private word with me. Behind the closed door, she told me that I seemed to have no idea how to use a secretary, and that she would teach me how.

She said that from now on, she'd handle many personal and business chores for me. She'd obtain office supplies for me. She'd take over the contacting of people I wished to have attend meetings I called. She'd write conference reports for me, based on my notes. I wouldn't even have to take the time to dictate them to her. By the end of our meeting, my secretary had taught me how to free up about 15 percent of my time, time that I ought to have been devoting to increasing company profits rather than tackling no-brain errands, tasks that can be handled by assistants, receptionists, people in the company with the time and ability to fill in the gaps between clerical tasks and answering telephone calls.

My production shot up dramatically after that talk. It's a good thing my secretary taught me about delegating. I certainly had never learned how to do it or encountered the need for it during college or during my early years in the world of business.

Once I had the tool of delegating, I was able to get down to the business of business and leave behind the time-consuming details. I was also able to delegate work to far more people than just my own secretary. As a result, my output shot up, my division became ultraproductive and successful, my company enjoyed increased profits from better time utilization, and I was rewarded with a raise, a promotion, a larger staff, more responsibility, and even a window office.

Let's hear it for my secretary. Yea, Doris!

Looking back at it now, I realize that although I thoroughly enjoyed the rewards, the raise really belonged to my secretary for teaching me what was important to do and what was important to delegate. Because she had served as secretary to one of our company's top executives before working for me, she had learned the difference between business necessities and business game playing. Bless her for passing that wisdom on to me. It made a permanent impact on my life because it freed up a lot more time for me to do what I was hired to do.

I now am able to work without a secretary, delegating all details to people who earn a living attending to such details, and probably love their work.

DEFUSING THE INFORMATION EXPLOSION

In Chapter Five is information you'll want to have about another tool that can save you weeks and weeks, maybe even months and months—a tool that can fill your life with ninety-minute hours. It's a simple tool, one with which you're undoubtedly familiar: the combination of an audiocassette player and a wide selection of audiotapes.

To succeed in business, in relationships, in self-improvement, and in the increasingly complex world of today, information is a powerful ally. There are many ways to obtain that information. Some people take courses. Others attend seminars. Others read books. Some pick it up from television. And still others spend the time and money for professional consultants. All of those are fruitful sources of information. And I believe it is far better to spend time gathering important data than to save time by ignoring the data altogether.

But best of all is to save time *and* gather the data. Having your data cake and eating it too is simple with audiotapes. Increasing amounts of high-potency business information is being served up on audiotapes these days. These tapes may be heard while driving, riding, flying, bathing, relaxing, waiting, and doing a host of other things that free your mind to soak up knowledge. With these tapes more minutes can be crowded into every hour; even more than ninety minutes.

Recently, while on a six-hour cross-country flight, I picked up six hours of information through audiotapes and my trusty personal stereo. I owned my first minihead-set in 1963, twenty years before they became standard equipment for music lovers and knowledge lovers.

By my own reckoning, I've saved over a year of classroom study, probably even more, by my incorporation of this technology into my life. I know people who have gathered this same information via other, more time-consuming means. I applaud their quest for knowledge but not their squandering of time. Some of these people are total taffypockets when it comes to saving money. But they seem to be spendthrifts when it comes to time. I hope your value system is on straight so that you don't waste away so valuable a resource.

That's the good news about audiotapes. The better news is that speed-listening devices, available for less than one hundred dollars, enable you to hear these tapes in half their normal running time. More about that in Chapter Five.

MAGIC AND THE NINETY-MINUTE HOUR

Chapter Six will prove beyond doubt that subliminal suggestion is one more tool in your potential arsenal of timesaving measures now available to the general public. All you've got to do to take advantage of this brilliant breakthrough in communication is to listen to pleasant music or the sound of gentle waves or even your radio while you're doing something that's really important. While your conscious mind is focusing on the task on hand, your unconscious is learning new skills, breaking old habits, absorbing powerful information, making you a more informed individual.

By means of the magic of subliminal suggestion—and it really does come across as magic, especially after you see how incredibly effective it is—you can lose weight, become a better salesperson, improve your vocabulary, give up cigarettes, learn how to learn, start on the road to financial independence, jazz up your sex life, develop techniques to increase your memory, even become a more astute investor. All this information can come to you with absolutely no drain on your time. In fact, the subliminal tape producers actually suggest that you do something else while hearing the tapes. Active listening is not required.

Sound a bit like hocus-pocus? It sure did to me—until I tried it and saw how it worked. Why it works and how it can work for you will be covered clearly and completely as you continue reading.

TV AS A TIMESAVER

As you'll see in Chapter Seven, a huge appliance like a television set and companion videocassette recorder may also lead to ninety-minute hours. Not only can you play subliminal videotapes with those tools, learning as you are entertained, but you can also use the VCR to cut down on the amount of time you spend watching TV—without cutting down on the enjoyment one iota.

In Chapter Seven, you'll also see how you can tap into outer space with your own earth station to increase your storehouse of information without infringing on your valuable time.

COMPUTERS: CREATED TO SAVE TIME

It will probably come as no surprise that a chapter will be dedicated to heralding the computer as a natural-born timesaver, though computers are neither natural nor born. Still, they do a bang-up job of loading your life with ninety-minute hours. They can save work time as well as communicating time. Such efficiency ought to be rewarded with a chapter of its very own, Chapter Eight.

PHONING FROM THE FREEWAY

Same for car telephones—see Chapter Nine. I don't think you'll have to stretch your imagination very far to recognize how a car phone can save time for you. But you might be surprised to learn just how *much* time one can save for you—or if not for you, for some time-cherishing friends of yours.

You'll get some idea of the importance (and status) some people are placing on these portable timesavers when you hear the apocryphal tale of the man who called his buddy on his buddy's car phone the other day. In the middle of their conversation, the man was told he'd have to hang up. When he asked the reason for the abrupt end to the call, his buddy informed him that his other car phone was ringing. You may not go that far, but if you go half that far, you'll be well on your way to mastering time extension.

REMOVING WASTED TIME FROM WAITING TIME

You'll also discover, in Chapter Ten, methods for transforming pesky waiting time into valuable accomplishing time. Whoever heard of waiting being a timesaving tool? You soon will. In fact, all of Chapter Ten will be devoted to changing your attitude about waiting. I doubt if I'll ever get you to relish waiting, but once you see how waiting can add extra minutes to your hours, I'll bet you'll consider it in a slightly more favorable light than you do now.

Part of this chapter will be devoted to several timesaving tools you probably own right now. I'm talking about a ballpoint pen and a small pad of paper. I'm also talking about something you may not yet own, but should: a microcassette player-recorder, the kind that easily slips into a pocket or purse. This type of miniature technology leads to many a ninety-minute hour.

EXTRA TIME FROM WITHIN YOU

You'll master the talent of time extension still further when you tap into an even more potent tool for creating ninety-minute hours. The tool to which I refer is your unconscious mind—the 90 percent of your brain that many scientists tell us is currently untapped. In Chapter Eleven you'll learn methods for enlisting your unconscious mind in your quest for ninety-minute hours. You'll be informed of special techniques that allow you to put that force on your side. You'll also learn of a few behavioral changes you'd be wise to adopt to extend time for yourself at home, at work, and between the two.

In addition, there are several keys to obtaining ninety-minute hours. Both deserve a chapter of their own; both get those chapters as the book comes to an end—bringing you to a new beginning, the brink of a life filled with the benefits of a multitude of ninety-minute hours.

You've seen how the ninety-minute hour works so well for Tom Dennison. You've seen how sixty-minute hours shortchange Stan Danton. Now, you're about to see how ninety-minute hours can fit comfortably and advantageously into *your* life.

2

How to Put Ninety Minutes into Your Hours

By making timesaving behavior part of your daily routine, and by utilizing as many timesaving technologies as possible, you can begin to accomplish two things at once. Things that used to take one hour each, a total of two hours of your life, can be handled simultaneously, in a total of one hour of your life. By dealing with life's and work's tasks in this manner, you will be giving yourself the gift of extra time. A conscientious practitioner of ninety-minute-hour concepts can even double many of his or her waking hours, gaining twice the opportunities of existence allowed by the old rules of time.

Now you and I both know that this book is not going to help everyone live two lifetimes in one—though that's a highly intriguing goal. But we will be able to stretch the number of experiences we embark upon and embrace. Time-oblivious people can use this extra time to dramatically improve their lot in life—if they feel it needs improving.

The overall result can average out as high as 50 percent more time during waking hours for each wise time investor. That's almost like adding thirty minutes to each waking hour. Thus we have access to the ninety-minute hour. You won't be able to verify the extra time with a stopwatch, but you will be able to vouch for its effectiveness with a consideration of your accomplishments, whatever they may be.

You'll also notice an abundance of new opportunities. Time has a wonderful way of exposing them to you. You can use these opportunities any way you wish—to accomplish personal goals, to achieve company objectives, to experiment with new activities, to become involved in a recreational pursuit, to broaden your education, or to loll about any way you choose amid the comforts of leisure time. My list is partial. You can make a longer one if you just let your mind wander a bit and think of what you might do with all that extra time.

A life rich in ninety-minute hours is the approximate mathematical equivalent of a 111-year life span—based upon a current life expectancy at birth of 74 years. Maybe you aren't interested in living the mathematical equivalent of 111 years. Could be you don't give a hoot about adding extra minutes to any one of your hours. But if that's the way you feel about time, you're part of a shrinking minority. Current research from the Daniel Yankelovich organization indicates that people are more aware of time now than at any other point in history. In addition, they place a greater value than ever on time.

The research explains this time consciousness as a result of the marked increase in the number of working women. In the past women handled all household chores, leaving men with much valueless time—time they could just waste away if they wished. But now that women are in the work force in such numbers, they don't have the time to do the majority of the housework. So men have less valueless time—less time open to any option they select. There's work to be done at home, and somebody's got to attend to it. How about you over there on the couch reading the newspaper? The net result is that everyone's time has increased in value. And that is precisely why we are more aware of time, more appreciative of its worth. That is why ninety-minute hours make more sense than ever.

Fortunately, the technology is here to make ninety-minute hours possible. Psychologists are helping us understand human behavior more clearly these days, so we know how to use our mental prowess, without even trying hard, to get what we want.

IF YOU EXPERIENCE STRESS, YOU'RE DOING IT WRONG

Don't get the idea for one nanosecond that a life filled with ninety-minute hours is a life filled with hypertension. In fact, the opposite is true. Many people, especially executives, feel that they are under too much time pressure as it is. Ninety-minute hours can relieve the pressure, reduce the stress. If these extended-length hours increase anxiety one iota, someone is probably forgetting part of the process of creating them.

That process includes planning. Without planning, pressure situations can arise. With planning, they are avoided from the get-go.

It is true that enlightened management will quickly recognize the myriad of production and financial advantages to their firm if both they and their employees practice the principles that create ninety-minute hours. And it would be foolhardy not to foresee that some managers will be tempted to abuse the privileges granted by extra time. Some will heap unreasonable chunks of work on the desks of their suddenly superefficient employees. But that is missing the point of the ninety-minute hour. It is abusing a good thing. Accomplishing extra work is one thing. Asking for one hundred minutes' worth of effort in ninety minutes is something altogether different.

The way to avoid stress—and this book is not about gaining extra time at the expense of enjoying life—is planning. This does not refer to standard plain-vanilla planning. It refers instead to a special type of planning based on the concept of *time extension.* You've heard of time management. You're familiar with time organization. Now you are being offered the benefits of time extension.

Time-extension planning will take you beyond time management by giving you more time to manage. That means you'll have more minutes to live and accomplish, so you're well-advised to plan how you'll use them.

You'll still have to organize and manage time, just as in the days before ninety-minute hours. Time management, or time organization, is planning the utilization of your time. If you're sophisticated about it, you practice time prioritization too. That kind of planning helps you accomplish the more important tasks first and the less important tasks last, so that if you can't get to all of your work, at least it's the less important work that gets the short end of your time stick. The important work is handled first because you gave it a high priority. That's always a smart decision and makes you a more efficient person.

Prioritization is in the spirit of the ninety-minute hour, but you can improve your efficiency even more through time extension. Time extension is the planning of work activities that you can combine or delegate so as to accomplish the maximum amount of work in the minimum amount of time. It involves organizing your time, to be sure. But it also involves thinking about the nature of the tasks. Can some be combined with others? Can some be avoided by you but nevertheless accomplished? Ninety-minute-hour thinking helps you see that more work can be combined with other work than you may have thought possible. It helps you realize which work demands your attention and no one else's. It also helps you realize which work only requires your touch or direction, but not your time.

HOW MEETINGS EAT INTO YOUR NINETY-MINUTE HOURS

There is no question that meetings are necessary to conduct business. There is also no question that too many people spend too much time in meetings. This is especially grave when meetings erode the time of the top echelon of a company.

The numbers show how much time the average chief executive spends in meetings each week:

HOURS IN MEETINGS	PERCENT OF EXECUTIVES
0–9	20
10–19	37
20–29	25
30 or more	18

Many of those top executives could have learned the meeting highlights from a conference report. There is little likelihood that their presence was necessary at every meeting. You can see that cutting their meeting time could mean a substantial increase in the time of many company executives, freeing them up to attend to more important matters.

Before attending a meeting, ask yourself:

- Can I get the main points of the meeting from a conference report?
- Do I have to attend in order to contribute?
- Can I participate in writing?
- What can I do to streamline the meeting?
- Can I help the company more by doing something else?

MISGUIDED ENERGY UNDERMINES NINETY-MINUTE HOURS

It is basic human nature to spend time doing the things you're good at doing—the basis of the Peter Principle.

That principle states that people rise to the level of their incompetence, becoming recognized and promoted on the basis of their work at tasks at which they were competent, but eventually getting in over their heads.

Spending the majority of your time at tasks that you do well is laudable. But if greater benefits would result from your doing something else, the misdirected devotion of your time could be working to the detriment of your business, your family, or your life. The idea is to spend time doing the things you've got to do. Of course you have to be good at doing them. But I implore you not to do things that can be done by someone less important to the success of your company or your business than you are.

SLOWPOKES AND THE NINETY-MINUTE HOUR

I realize that many people intentionally work slowly. The ratio of time involved in work to time available to work is usually 0.6—which means 4.8 hours of an eight-hour day are spent in actual work. I also know that it takes about three overtime hours to produce two standard hours' worth of work. For truly heavy work, it takes two hours for each hour's worth of output.

White-collar work efficiency isn't any award winner either. The average white-collar employee spends three weeks a year socializing at work. Two-thirds of one thousand personnel executives believe that personnel put in a good half hour a day talking about movies, friends, lovers, and assorted subjects of gossip. One-fourth of the executives claim those chat breaks consume a full hour.

I don't want to throw my monkey wrench into human nature. But I do urge you to put more minutes into as many hours as possible so that you can accomplish more even while continuing to work slowly.

As you become more efficient, be on guard for symptoms that indicate a flaw in your planning:

- Additional stress as a result of adding more minutes to your hours

- The feeling that you are being transformed into a working machine

- Time spent twiddling your thumbs because you finished all your tasks ahead of schedule

- The sense that you're being dehumanized because you are enlisting the aid of electronic technologies such as audiotapes, microcassette recorders, VCRs, computers, and car phones

- Psychosaturation with timesaving behavior and technology as you experience the benefits of work delegation, subliminal suggestion, and access to your unconscious mind

- Any association of the idea of ninety-minute hours with some form of pressure

- A pace you've set for yourself with which you cannot feel totally comfortable

- Too much free time—becoming a bum because you accomplish so much in so little time

- The temptation to use your extra time solely to do more work instead of striving for a balanced life, thereby risking the disease of workaholism

USING TIME EXTENSION TO ACCOMPLISH YOUR GOALS

If you're to be a person blessed with a plethora of minutes in your hours, you've got to start out with the clear aim of creating time. If you do, it won't take long for you to comprehend that the key to creating time is saving time. Once that idea is firmly embedded in your mind, as evidenced by your streamlined behavior, you'll begin to be a time-conscious person. But first you've got to get into the practice and mind-set of time extension. You've got to think in terms of what time you can extend.

The planning that is necessary to create extra minutes for your hours begins with three simple questions:

1. What tasks do I want to accomplish?
2. What tasks can I combine with other tasks and still accomplish?
3. What tasks can I delegate to others?

After you ask and answer these questions, you can begin to use time extension to accomplish your goals. At first you might have to work at it. But the fruits will be so rewarding that the process will soon become automatic. You'll begin to think, in time-extension terms, of both work tasks and living tasks. You'll be given more time than ever to learn. You'll develop a healthy respect, if not a profound reverence, for time—especially that time you may have tended to ignore in the past: waiting time, travel time, time spent unable to sleep, meeting time, time on the telephone, and time engaged in idle chitchat. You'll start to become appalled, especially if you're the boss, at the notion that waiting for work takes up to 15 percent of the workday of company clerical employees.

Waiting for work is no less a time bandit than working too long. In addition to being against our deepest nature, working too long is counterproductive. Bernard Casey, a labor economist at London's Policy Studies Institute, says, "There's a diminishing return, the longer you work. Longer hours do not necessarily mean more production."

Japanese manufacturing companies have calculated just when longer hours start to result in lower productivity; they arrived at 2,300 working hours a year. In West Germany the figure is 1,700 hours. Stateside, after deducting holidays and vacations, we put in 1,900 hours. Typical of workaholic executives, the Japanese felt that 2,300 hours weren't enough for them.

Jean-Pierre Lehmann teaches at a business school in Fontainebleau, France, and realizes that for Europeans to keep up with their Japanese competitors, they need to work harder. But he adds, "The point isn't to spend more time in drudgery. It is to use time more productively."

Once you're a time-conscious individual, you'll almost automatically become more productive in your work, harvesting a healthy crop of extra time in the process. Soon you'll recognize the choices that are yours when you create that extra time. You can use it in one or more ways: (a) for productivity; (b) for learning; (c) for leisure. My recommendation: use it for all three. But know up-front just what you'll do with the time you create. If you don't, it will be difficult to create extra time. The motivation just won't be there.

Combining tasks by handling them simultaneously will create extra time. Whether you do them simultaneously, handling both by yourself, or you handle one and delegate the other makes little difference. The net result is still extra time.

Practicing time extension will enable you to reduce the time necessary for many activities while creating extra minutes for your hours at work, at home, and traveling between the two.

Because I work from home, a privilege that I disparage in no way, I am deprived of valuable commuting time that can be used, by employing time-extension technologies such as listening to instructional audiotapes, to enhance my value as an employee or a human being. I don't pine for a long commute, but I know that some of my friends use their commuting time in a variety of ingenious ways. I also know of others who waste that time in a variety of ingenious ways. Some actually cherish their commute to and from work; others moan about it.

As an innocent noncommuter, I see how the cherishers are enjoying a vastly higher degree of business and personal success than the moaners. But I judge not, for if I were car bound or train bound for two hours a day, I might turn into a moaner in spite of the working and learning opportunities afforded by a long commute.

DIPPING YOUR TOES INTO THE TIME-EXTENSION WATERS

In addition to the many time-creating approaches that I've mentioned so far, there is also a healthy collection of time-compressing skills that can be immensely valuable. These may not be dramatic methods of putting ninety minutes into your hours, but they do create extra time. And you may be using some of them now. If so, you have already tested the waters of time extension.

Shorthand and speed writing are methods of time extension. Speed reading is another. Like it or not, we are well into the Age of Information. And sociologists predict that eventually the world will be composed of two classes: the information society and the no-information society.

If you're a specialist in speed reading, you'll have better credentials to enjoy the advantages of membership in the former society. Speed reading will give you the benefits of time extension by creating time for you that does not exist for those who read at the standard 250 words per minute.

You can also reap those benefits by gaining the maximum amount of information via newsletters rather than books or magazines. The proliferation of newsletters on virtually any topic is a response to the higher value people are placing on their time. They want their information, but they want it to be presented briefly. You can imagine how much time is created by a speed reader of newsletters. That person gains the maximum amount of data in the minimum amount of time. It doesn't sound very stressful to me. But it does sound practical. And I'm only talking about commonplace methods of saving time— speed reading and newsletters—not about the uncommonplace method of combining tasks.

One of the most common methods of saving time is by typing rather than writing in longhand. Many people who must write things haven't yet even moved into the age of the typewriter, let alone the word processor. They're still using their quill, though some have advanced to ballpoint technology. To these people, I say: If you *must* write, dictate or type.

Only recently has America embraced the concept of

time organization. Only recently have we learned to make lists of our activities and assign a time to when we will handle them. Only recently has the concept of time prioritization come into the public awareness. Only recently have ninety-minute hours become possible.

If there is any homework to accomplish in order to put a full complement of ninety minutes into your hours, that homework is to begin managing your time. I don't, however, mean you should invest in one of those ornate, complex, high-tech, multicomponent systems of time organization. It seems to me some of them require as much time to maintain as they ought to be saving for you.

YOUR CAR AND THE NINETY-MINUTE HOUR

I am not going to recommend that you drive 130 miles per hour in a 65-mile-per-hour zone to save time, but if you could do it legally, you'd be covering the same amount of distance in half the time. And you'd be discovering the basic idea behind the ninety-minute hour. I'm not even going to recommend that you have an audiocassette player installed in your car—though I will make that recommendation in Chapter Five—so you can take advantage of the galaxy of audiotapes now available.

Instead, I'm going to ask you to use your car as an example of how you can put more minutes into your hour. For instance, it makes a whole lot of sense to have your car serviced at a garage near the airport while you are flying off on a trip, however brief. This saves you the time of driving your car to its usual service facility, saves you the time of waiting for the servicing to be accomplished, and saves you the time of transporting yourself to the service facility to pick up your car. No wonder there's a garage right near the Cleveland airport that actively solicits air travelers with this pitch. No wonder the garage is so profitable.

Time-conscious people save a car's tune-ups, oil changes, and lube jobs for when repairs or replacement parts are needed. One trip to the garage does the work of two, and sometimes three.

Putting ninety minutes into your hours begins with orienting your mind to seeing the abundant opportunities for the creation of extra time. Once you've got down pat the art of putting ninety minutes into your hours, you'll naturally want to put as many ninety-minute hours as possible into your life—at work and away from work.

In the spirit of accomplishing that as soon as possible, let's get down to the details of how it should be done.

3

How to Put Ninety-Minute Hours into Your Life

There is an endless supply of work. There is not an endless supply of time—at least not for you and me there isn't. So regardless of when you complete the work assigned to or undertaken by you, remember that someone will be able to figure out more work for you. That's why it's so important to schedule in free time. If your boss can't find something else for you to do, I'll bet you'll have no problem coming up with tasks.

In your advance planning, see which tasks you can combine, which you can delegate, which you can streamline, and which you can skip entirely. If you get nothing else out of this chapter, get this idea down pat: One of the secrets of living a life filled with ninety-minute hours is to prime your unconscious mind to save time, to spend time wisely, to extend time, and to take complete control of time. Merely by understanding the principle of saving versus wasting time, your unconscious mind will become primed for the crucial chore of advance planning.

Living a life replete with ninety-minute hours will require that you make advance planning second nature—just like driving a car. You can probably drive your car ninety miles per hour without a second thought, except for a glance in the rearview mirror to be sure John Law isn't following you. So it should be in living a life of ninety-minute hours. The more you do it, the easier it will become. Automatically, you'll know what tasks must be undertaken with your full concentration and what tasks can be combined or delegated. When you can unconsciously assign your tasks, both business and personal, to the handle-myself, combine-with-others, or delegate-away categories, you'll be primed to add the maximum number of ninety-minute hours to your life.

The advance planning means not only deciding which tasks to handle, to combine, and to delegate but also deciding exactly what must be done, who ought to do it, when it must be done, and how can it be done most efficiently. Once you make those items part of your planning, you can move on to other methods of adding time to your life.

HOW TO USE PROCRASTINATION TO GAIN NINETY-MINUTE HOURS

The way to employ procrastination is not necessarily to avoid it but to control it. As you probably know by sheer instinct, procrastination is the enemy of the ninety-minute hour. I handle procrastination by allowing for it, engaging in it, expecting it, welcoming it, then getting it over with. We all know that procrastination is a natural, human, and healthy trait. So don't fight it, incorporate it into your life.

For example, I set aside a full hour each day simply to put off the important work that I must do. It's strange to admit, in a book about saving time, that I intentionally waste time, but believe me, I do waste it—in some of the most creative and ridiculous ways possible. Yet unless I gave myself that hour to piddle away, I don't think I'd have the mental stamina to accomplish as much as I do during the rest of the day. Once my procrastination is over, I generally work a straight eight hours: no lunch, no coffee breaks, no phone conversations, no low-priority work, and no socializing.

Procrastination primes your unconscious mind to get down to the tasks at hand. It builds up a sense of guilt, which you can only conquer with solid and constructive work. To procrastinate properly, give each bout a beginning, a middle, and an end. With such control, procrastination will be transformed from an enemy to a friend. It will create a sense of unconscious obligation for you to complete the tasks you have promised you'd complete.

The way to end procrastination is obvious: Start doing something else. The key is in starting. As far as I'm concerned, the act of writing a book is only a two-part process: starting and completing.

Starting is a bit harder than completing, but takes a far shorter time. In the case of a book, it's only writing one page. The completing phase might mean writing two hundred more pages. Yet both are equal in importance. Deep down, know that starting will invariably lead to completing. I do all of my procrastinating in the morning. I am far too busy in the afternoons to procrastinate. That time is solely devoted to completing what I have started. If ever I'm tempted to procrastinate in the afternoon, I need only think back to the morning, when I proved myself a world-class procrastinator; then I hurl myself into the task at hand. The net results of controlled procrastination are extra accomplishments, extra free time, no guilt whatsoever, freedom from uncertainty or stress, and a compatible relationship with the enemy—procrastination.

Do I recommend my method of procrastination to you? Of course not. But I do recommend that you deal with procrastination by accepting it and treating it in a fashion tailored to your needs. The word to remember is *control.*

HOW TO DEAL WITH INTERRUPTIONS IN YOUR NINETY-MINUTE HOURS

Another method of gaining more minutes in your hours is remembering how sacred each minute is and steeling yourself against anything that would erode those minutes. This awareness involves recognizing that interruptions will happen—but doing everything in your power to prevent them.

Interruptions are preventable in at least five ways:

1. Make it known to all who deal with you that certain times of the day are not interruptible. My friends, associates, clients, and family are so familiar with my schedule that they wouldn't dream of trying to contact me during working hours. They know there is plenty of time to con-

nect with me, and they know when that time is. If they absolutely must get in touch with me during work, they keep the conversation down to a minute or less. I can handle interruptions of less than sixty seconds.

2. Tell strangers who interrupt that you are very busy now and will get back to them as soon as your task is over. Speak politely, note their number, then return to your chores. That usually takes under a minute. Even the most high-energy telephone salespeople respect your request for peace to continue the task at which you are working. Do the same for associates with a propensity for popping into your office.

3. If the interruption can be fielded in less than a minute or so, allow it to happen. If it will take longer, put it off. You can fairly well ascertain the length of the interruption by asking, "How long will this take?"

4. Set up times during which you encourage interruptions. Let your staff, friends, or associates know of these times, and invite them to interrupt you like crazy during them. People—even kids—quickly learn the difference between such times and noninterruption times, and they act accordingly. I've seen this system work in a crowded office as well as in a private home. It works because people like clearly defined boundaries and tend to respect the most clearly defined.

5. Delegate all interruptions to someone or something else. Let them be handled by an associate, even an answering service or answering machine. When it is time for you to work without interruption, the delegating begins.

Once you analyze the nature of your interruptions and the people who do the interrupting, you'll be able to deal even better with these enemies of the ninety-minute hour. You'll know why you are being interrupted, and you'll be able to eliminate interruptions by acting accordingly—perhaps issuing a memo that answers the most frequently asked questions. You can talk to the 20 percent of the people who are most likely doing 80 percent of the interrupting. When you talk to them, be compassionate, sympathetic, helpful, and sensitive to their needs—but also be clear that they are getting in the way of your efficiency.

Experts claim that merely by avoiding procrastination and interruptions, plus planning your day, you can save two hours a day. If you do these things as well as take advantage of new technologies, you double that savings. And then you can even do a bit of procrastinating, if you allow for it.

THE IMPORTANCE OF MAKING A LOG—IF ONLY ONCE IN YOUR LIFE

When I quit smoking I did it by first making a log of the cigarettes I smoked. I had never given the matter much thought before, but when I made a log, I realized at least two things that floored me. First, I was smoking a full pack from the time I got to the office, where I had my first cigarette of the day, until 4:00, about an hour before closing time. Second, I rarely smoked more than one or two cigarettes after dinner. Had I been asked, before making the log, when I did my smoking, I would have said that I spread it out through the day. But by making the log I was able to recognize my pattern, take control over it, and end it. It's much easier to defeat an opponent you can see than one who is invisible.

Your first task in your quest for ninety-minute hours should be to make a log of the activities you carry out in a day. Just write the activities down after you do them so you can look back and get a feel for how you spend your time. I suggest making a log for one full seven-day week.

The idea is not to change your behavior, but to teach you about yourself. When you look over the one-week log, you'll have a much clearer idea of how you spend your time, which time you are wasting, which you are using to good advantage, and which can be streamlined. Like my cigarette log, I'd wager that your log will surprise you. I'm not sure exactly what will surprise you; I'm only sure that you'll gain insights from the perspective of your log.

In making the log, don't attempt to save time, accomplish more, manage time, avoid procrastination, or do anything noble like that. Instead, simply record the *expenditure* of your time during a normal week. And remember to keep the log for both your working and your nonworking hours. To gain extra minutes for your ninety-minute hours, you'll need to find them wherever they may be.

SLEEP MY CHILD, AND PEACE ATTEND THEE, ALL THROUGH THE NIGHT

For the purposes of this book, I encourage you to get eight hours of sleep per night, even more when you're ill. I realize there are several people out there who eliminate one or more hours of sleep per night to accomplish more work per day. Still, although I am not suggesting that you sleep nine or ten hours a night, I am against cutting out any of that valuable time during which your body and your mind are receiving their fair share of rejuvenation.

I am all for you sleeping sixty-minute hours—eight per night. And I do not recommend you give up that sleep for anything—with rare exceptions. I do, however, recommend that you put sleeplessness to work for you. Eighty percent of people over the age of twenty-five experience sleep problems. The older you get, the more trouble you will have staying asleep. Some people are thrown out of kilter by lack of sleep.

But that is not a problem if you consider sleeplessness to be your ally. Instead of lying there in the dark, bemoaning your inability to go to sleep or stay asleep or go back to sleep, *do* something. Amass a list of tasks that can be undertaken during periods of sleeplessness. Naturally, none of these will be high-priority tasks, because sleeplessness is relatively unpredictable. But many will be important tasks that you just can't get around to during your daylight hours. With such a list, you will end up welcoming sleeplessness because it will provide so many opportunities for you. Some of the extra minutes for your ninety-minute hours will come from these periods of sleeplessness. And I highly recommend that you put them to use.

You might want to write a letter, clean your files, or dictate a memo. Maybe you'll be clearheaded enough to make an important decision. After all, during sleeplessness you are in closer touch with your unconscious mind than during full wakefulness. Your mind's abilities might impress and surprise you if you put them to work, rather than counting sheep, eating sleeping pills, or staring at the ceiling.

You'll recognize several tasks that are perfect for sleepless periods. Keep your list of them next to your bed. Also keep a pad of paper and a pen there, along with memos, books, magazines, or newsletters that require your attention.

I wish I could honestly tell you that you can learn while sleeping by listening to a sleep-learning tape. But alas, I have yet to hear of a study proving that sleep learning works.

THE UNSUNG TIMESAVERS

There are many nonglamorous timesavers available to you. Their hidden value is their ability to save you small segments of minutes rather than giant chunks of hours.

Catalogs and stores are replete with merchandise that, among other things, saves time for you. An ubiquitous example is the array of timers on the market today. They'll turn on your lights, turn off your lights, start your oven, turn on the stove, water your lawn, and do a host of other timesaving chores for you. They are revered by devotees of the ninety-minute hour. Individually, they don't save that much time. But when you consider all of their applications over the course of a month, you begin to realize that they save several hours for you—hours that can be spent doing anything you wish other than turning appliances on and off.

LETTING YOUR TELEPHONE SAVE TIME FOR YOU

Although I will go into heavy-duty detail about the esoterica of timesaving, such as subliminal suggestion, in Chapter Six, I do want to direct your attention now to the simplicity of such commonplace timesaving devices as a long cord on your telephone, or better yet, a cordless telephone. Best of all are the cordless phones with built-in mouthpieces. Freedom for your legs and your hands.

With the long cord or the lack of a cord, you can attend to certain details while on the phone. At home, you can continue to cook or clean or whatever you were doing when the phone rang. At the office you can file or categorize or give a cursory reading to business matter. Some conscientious timesavers add a shoulder attachment to their phone for that purpose.

THE NINETY-MINUTE HOUR AND THE REST OF THE WORLD

Our whole world is joining in the quest to save time. Automated teller machines at banks certainly save time over waiting in long lines for available tellers. Now that people are becoming more adept at using ATMs, the lines move faster than ever. Banking by mail is one more time-saver. I once went for slightly over five years without setting foot inside my bank. I continue to bank by mail and continue to wonder about those long-line lovers who don't.

I haven't waited in a line to see a movie in years and years. Either my wife and I show up just as the movie is about to start and the line has gone inside—or else we wait to rent the videocassette. We didn't even have to wait to see *Star Wars* because we got there right at show time. Yet the newspapers were filled with stories about the length of the lines. No question; that day the Force was with us.

More line avoidance comes when you use curbside check-in at airports rather than waiting in the interminable lines of travelers inside. I manage to snag a window seat right up near the front of the plane simply by preselecting my seat—one more timesaver.

Right here and now, I feel it my duty to call your attention to one of the greatest masters of time usage in our lifetime: Alan Lakein, author of the fabulous-selling *How to Get Control of Your Time and Your Life.* Although the book was published in 1973, its truths are just as relevant today as when the good Mr. Lakein wrote of them.

One of his most famous contributions to the saving of time is his "80/20" rule. The rule states, "If all items are arranged in order of value, 80 percent of the value would come from only 20 percent of the items, while the remaining 20 percent of the value would come from 80 percent of the items." Although this rule is not hard and fast, it is true about 80 percent of the time.

He goes on to suggest that if you make a list of ten activities, doing only two of them will yield 80 percent of the value of the whole list. He recommends that you identify these two items, put them up at the top of your priority list, which he terms your "A" list, then do them.

To add to the credibility of his list, Lakein tells us of these other 80/20 examples, most of which you'll instantly recognize as being true:

- 80 percent of sales come from 20 percent of the customers.

- 80 percent of production is in 20 percent of the product line.

- 80 percent of sick leave is taken by 20 percent of employees.

- 80 percent of file usage is in 20 percent of files.

- 80 percent of washing is on 20 percent of the wardrobe.

- 80 percent of TV time is spent on 20 percent of the programs.

- 80 percent of reading time is for 20 percent of the newspaper.

- 80 percent of phone calls come from 20 percent of callers.

- 80 percent of sales are made by 20 percent of the salespeople.

It is your job, and not a very difficult one, to sort out among your life's tasks which items fall into the valuable 20 percent and which belong in the less valuable 80 percent. An unconscious that is primed to save and extend time for you will find this a simple matter. But an unconscious that lacks this priming will not differentiate between the two. Just remember: All tasks are not created equal.

One final tribute to Mr. Lakein must be paid here. It is committing to memory what he terms Lakein's question—"What is the best use of my time right now?"

The answer ought to be obvious all the time. But the truth is that it is not obvious. In fact, the usual answer is "I don't know." However, once your unconscious becomes aware of the value of your time, the limited amount of time you have here on planet Earth, and the opportunities to extend time, I think you'll know the right answer. At least you will 80 percent of the time.

You do not exist in a vacuum and are not the only occupant of the universe; there are others around, about 5 billion others at latest count, who just might be able to help you save and extend even more time. Let's talk about some of these people in the next chapter.

PART TWO
THE TOOLS

4

Mastering the Art
of Delegating

Everyone knows about delegating, but not too many people do it, and fewer still do it well. The books on time management strongly recommend that you delegate work, but they don't tell you how to do so, and chances are slight that you have experience at delegating. It's not easy to delegate intelligently, as is proven by the numerous workaholics foisted upon our society. Although delegation is a basic saver of time, it is not easily mastered. Yet mastery of it provides you with a significant tool of the ninety-minute hour.

To gain the countless benefits of the ninety-minute hour—benefits limited only by your own imagination—

requires that you utilize and master a variety of tools to accomplish your goals. So adroit should be your handling of these tools that you learn to consider them valuable allies. These allies come in many forms: as concepts, information, psychological techniques, perceptions, technological hardware—all ready to speed you toward satisfying your objectives for each day.

OBJECTIVES ARE NECESSARY

As one school of management has proposed, management is more easily and efficiently handled via a method called Management by Objectives. This results in streamlined, get-it-done meetings.

Here's how it happens: The meeting leader writes the objectives of the meeting on the blackboard. The leader then invites all in attendance to add to that list any objectives that may have been left out. This enables everyone at the meeting to bring up any issues that require discussion, attention, or both. The finished list is then discussed, item by item, with each item crossed out as responsibility is taken for it, or as it is permanently banished from earth.

Such meetings end up with an overall feeling of group satisfaction because all items are aired, all responsibilities assumed, all questions (in the best cases) answered. Several people leave with chores to attend to. Many successful businesses continue to conduct their meetings in this effective manner.

Your days, business or otherwise, might be effectively utilized if they too were aided and abetted by a list of objectives, furnished by you. This is more true for your workdays, since the idea does seem too organized for a nonworking day.

Each second, minute, hour, and day of your work time should be devoted to the accomplishment of your objectives. Effective work time is not focused on a particular meeting, proposal, task, production run, or personal encounter as much as it is on the satisfaction of clearly understood objectives, written on that blackboard in your head.

Job number one is to meet those objectives. Job number two is to meet them using the ninety-minute hour. It becomes obvious that when you have mastered the ninety-minute hour, you can meet more objectives.

THE EFFICIENCY OF NINETY-MINUTE-HOUR THINKING

Concentrate on the accomplishment of your objectives, but begin to apply ninety-minute-hour thinking: Is it necessary that you be the one doing the actual accomplishing? Don't forget that you are a multifaceted power— regardless of your situation at work or at home. Your ability to cause action, which is the way most objectives are attained, will determine the extent and effectiveness of your power—and your life.

Let's examine eight methods for exerting your power and attaining your objectives.

1. You can do the necessary physical labor with your body to cause the action that meets your objective. It takes time, and if you're good at it, you're certain the job is being done right.

2. You can do the necessary mental labor with your

mind to cause the action that meets your objective. It takes time, usually less time than physical labor, and if you're a solid thinker, you can be confident the job is being done right.

3. You can find a person on your level or just below and delegate the whole responsibility to that person—the mechanics, politics, elbow grease, details, or whatever it takes to achieve your objective. If you do your homework, that person will do the work as well as you would.

4. You can set up a system, using another person or people in your company, including your secretary, assistants, spouse, or kids for attaining your objectives, today and into the future. You've got to take the time, which might be well spent in this case, to train these people to do the work exactly as you wish it done—or better.

5. You may call in an outside individual or company to satisfy your objectives. If you go about selecting them properly, this company will specialize in the types of objectives with which you are presenting them, so you can be assured that they will handle your work responsibilities correctly.

6. You may hire an individual to take on these and similar objectives on a regular basis. This option may be open to you as an employee or as a self-employed person. Perhaps you can justify the person's income with the increase in income or free time his or her contribution of talent and time will mean to you. Once again, you must train the person to provide all the quality expected of *you*.

7. You can systematize the work by putting it on a computer program that will meet your objectives. If you're as good at programming as you are at attaining your objectives, you've got it made.

8. You can hire a computer programmer to systematize the work in a computer program. Ideally, you can

explain your needs clearly enough for the objectives to be consistently solved by a computer.

Ninety-minute-hour thinking strongly advises that you delegate as many of your objectives as possible—just as long as they are handled every bit as well as if you were handling them. It urges you to develop a frame of mind that guards your time and increases your effectiveness.

It's not hard to see the efficiency of leaning on all these eight sources of accomplishment. It is apparent that you, as an individual serving as two of these sources, compose only 25 percent of the potential objective-attaining force. It is also probable that right now you personally take the time to satisfy considerably more than 25 percent of your objectives.

THE DIFFICULTY OF DELEGATING WORK

People do work rather than delegate it for a variety of reasons. Sometimes they don't realize they can delegate it. Sometimes they don't understand the capabilities of modern computers. Sometimes they aren't aware of the economies inherent in delegating. Often they don't have anyone around to whom they'd entrust the work. Or they don't know of new services that exist solely to do the type of work they wish to delegate.

But primarily, their egos get in the way of their efficiency. They may feel there's someone around who could do the work, but they seriously doubt if it could be handled as well as they'd handle it. And they worry that if others successfully do the work, this may prove they are unnecessary.

There is little wonder that a wave of corporate cost cutting is sweeping across America. If our business community is in its right mind, the cutting will focus on the unimportant detail work in which their highly paid executives become involved. These details may provide a welcome break from making momentous decisions, but they're a waste of personpower, money, and shareholders' profits. Ivory-tower executives become involved in work that they could easily shuffle off to an assistant or one of ten others who would do the job with more cost efficiency.

Many secretaries are able to write letters that are every bit as readable, professional, and motivating as their bosses—but their bosses still persist in dictating. A good secretary needs only to be told the gist of the text, then to be left to create the right kind of letter. If the letters are repetitive, create a format letter that your secretary can use as a guide for future correspondence. The idea is that your job is probably not to write letters.

In my travels through corporate America, England, and France, I noticed some people who were absolute geniuses at delegation. The key was the utilization of their assistants. Their assistants attended meetings for them, did a fair share of traveling, handled any corporate gift shopping, attended to most correspondence, answered and instigated memos, made flurries of phone calls, even attended social functions in the boss's name. Proper recognition of an assistant's talents is one of the secrets of attaining the ninety-minute hour. Once you've recognized the talents of others, delegation is that much easier.

Spend time making a list of the kinds of work tasks that you can delegate. Although I'm confident that you can add to this list, let me give you fifteen as thought starters:

1. Sorting your mail
2. Opening your mail, when appropriate

3. Answering your mail, when feasible
4. Answering your telephone
5. Returning your phone calls when possible
6. Initiating your phone calls when possible
7. Reading your memos
8. Underscoring your memos, whenever possible
9. Responding to your memos, when appropriate
10. Initiating memos, signed by you or your assistant
11. Keeping your office supplies consistently stocked
12. Traveling for you, whenever possible and professional
13. Protecting you like crazy from taking work home
14. Holding meetings for you at which information must be imparted but your presence is not necessary
15. Staying alert for ways you can streamline your office life to improve your efficiency

Even if your assistant is young and/or relatively inexperienced, he or she probably does not lack intellect, organization, or ability. You'll have to do a certain amount of training and explaining before you delegate to a young assistant, but the time you spend on this preparation will be amply rewarded by the time you save—by propelling you that much closer to the ninety-minute hour.

WHAT YOU MUST DO TO PUT MORE MINUTES IN YOUR HOURS BY DELEGATING

It all starts with self-appraisal, knowledge of others, awareness of compatible services, and honesty. You've got to be able to look over your list of objectives for the day, then do all in your power to figure out to whom you can delegate the work. You've got to be candid about your own prowess, face up to your own expendability, admit to the talents of others, then delegate to the proper persons.

These people gain by learning. The company gains too, because more of your time is freed up for the primary tasks for which they hired you or for which you founded the company—probably not menial, delegatable work. And you gain because you have more free time to be truly productive while attending to necessities rather than doing menial tasks.

A LEAN TIME MACHINE

In pursuit of the ninety-minute hour, think of yourself as a lean time machine. Trim off all unnecessary fat. That fat comes in the form of delegatable work that you are not delegating. That fat is valuable time you are spending on tasks that others could be attending to for you. The more fat you trim, the leaner a working-and-earning machine you become.

As a result, you begin to meet your objectives without personally spending so much time doing so. You begin to use the power that has been yours all along. This enables you to enlarge and broaden your objectives—to make bigger, bolder plans. The ninety-minute hour gives you time to dare.

A GLIMPSE INTO THE NINETY-MINUTE HOURS IN MY HOUSE

If delegation works wonders in the office, how does it fare at home? I have learned the answer to that question firsthand.

Without quite knowing what we were doing, my wife

and I agreed to share our apartment with a live-in helper when our baby was born. I admit that I resisted the idea for about six months after Pat suggested it, worrying about the privacy we'd sacrifice. But we compromised, a face-saving way of saying my wife won the disputed point. We rented a small, cramped, three-bedroom apartment: one bedroom for us, one for daughter Amy, one for our live-in helper, Annie Mae.

Then I sat wondering whether I had traded my private life for a built-in baby-sitter. Now I sit congratulating myself for having the good intuition to go along with my wife's better intuition. A full twenty-five years have passed, a quarter of a century of living with a live-in helper. Amy has flown the coop, and we still have a live-in helper because of the immense amounts of time that she adds to our personal lives.

A word about finance: I write not here of fancy sums for full-time house help, but of fifty to seventy-five dollars per week for doing timesaving chores, freeing up valuable hours that give my wife and I the time to accomplish far more than we could without the paid help we have had over the years. We pay these wonderful people (usually students or part-time jobholders) $40 per week plus a $10 car allowance—they've got to have a car to live with us, so numerous are our errands. We also provide them with a room complete with bathroom, view, color TV, storage space, desk, free laundry facilities, postage stamps, three days off per week, seven evenings off per week, privacy, and all the food they want.

Rather than writing what we receive in return, I thought it would be more eloquent to print the list of chores for which our live-in helpers are responsible during the few hours they devote to achieving the daily objectives of our household.

When looking over the list, consider that it is the very essence of delegation, because it represents time that my wife and I do not spend attending to these details.

RUNNING THE HOUSE ON SEAVIEW DRIVE

Mornings
1. Get newspaper from top of driveway.
2. Be sure dog and cat water supply is ample.
3. Clean kitchen sink; neaten kitchen and guest bathroom.

Afternoons
1. Empty trash from all wastebaskets (replace white bags).
2. Take clothes to or pick up from dry cleaner.
3. Make beds.
4. Clean up kitchen counters and sink.
5. Empty dishwasher.
6. Clean up all four bathrooms.
7. Be sure no light bulbs need changing.
8. Be sure paper spools are filled, Kleenex replenished.
9. Fill cat bowls—food and water.
10. Feed dog.
11. Neaten up inside and outside the house, including driveway.
12. Get mail, put on Jay's desk.
13. Pay all bills as appropriate once a week.
14. Do weekly shopping.
15. Do the errands as necessary.

Early Evenings

1. Turn on fourteen lights just before dark when we're at home.
2. Turn on three lights just before dark when we're not at home.
3. Prepare dinner Monday, Tuesday, Wednesday, Thursday.
4. Clean up dinner table, do dishes.

Things Always to Keep Filled

1. Hummingbird feeder (except in winter)
2. Tea jar next to kitchen sink
3. Sugar pourer
4. Coffee jar in refrigerator (with fresh ground coffee)
5. Napkin holder on counter
6. Pepper grinder on counter (plus salt and pepper shakers)
7. Match holder in living room
8. Living-room plants (watered)
9. Water mug next to Jay's bed (fill each day)
10. Swimming pool (fill to top weekly)
11. Pet food supply
12. Orange juice pitcher in fridge
13. Iced tea pitcher in fridge
14. Coleus root system (water one time weekly: Tuesday)
15. Guest-room desk plant (water one time weekly: Tuesday)

By attending to these multiple and idiosyncratic details, our helper gives Pat and me the joys of much extended time, and a whole lot of ninety-minute hours. That list, now twenty-five years old and growing, elaborates the timesaving advantages of live-in power. If you checked each item listed, you'd recognize how much time is saved by never seeing the inside of a supermarket, drugstore, office supply store, bank, vet's office, hardware store, and more. You'd also recognize that live-in power can be altered to become secretary power, assistant power, or even kid power. There are many logical people to whom activities can be delegated and many logical reasons to do the delegating. It's simply a matter of breaking away from the puritan work ethic, ending senseless routines, and subordinating your ego to the needs of the moment—be they corporate or personal.

Every one of your delegating options can contribute to effective time extension. They all allow you to accomplish one of your objectives while someone else is accomplishing another. By now you know that this is the essence of the ninety-minute hour.

If you cannot delegate in work or nonwork situations, you cannot benefit from maximum time extension and the maximum amount of ninety-minute hours. If you can delegate, see how much more you can delegate. Think of delegating things you *never before dreamt* of letting go of.

A HORRIBLE FACT OF DELEGATION TO REALIZE

I sense that if you face this fact, you might be able to surmount the obstacle it presents: Most people lack the talent and inclination for delegation.

If you believe you belong in the category of nondelegators, do something about it. Break habits, even little ones. Delegate small matters until you are confident that you can delegate big ones. Wean yourself from self-dependence to dependence on others. Don't forget, the world is populated by millions who would gladly do the work you feel you must undertake.

Sure, they want to be paid for it. And sure, you'll have to do a fair amount of investigation before you know to whom you ought to be delegating. Still, the art and science of delegating is one of the staunchest allies of ninety-minute-hour devotees.

I want to build a bridge in your mind—a bridge connecting the delegating of work with ninety-minute hours. Cross it in the quest for more time, less stress, more control, less frustration, more accomplishments, less necessary expenditure of time and energy.

In no way should you think you can ask just anyone to do work that is within your domain. You've got to carefully check out the person to whom you are delegating. In the battle for extra minutes per hour, half the job is identifying the work that can and should be delegated. The other half is identifying the people who ought to be doing it.

TIPS TO HELP YOU DELEGATE WISELY

- Don't delegate to someone who won't do the work as well as you can do it.

- Recognize that every time you delegate successfully, you double your effectiveness.

- Don't delegate tasks to someone if you're not willing to train that person to do them excellently. And let that person set the terms, timetables, and objectives so he or she can measure how the work is going.

- Don't always tell the person to whom you are delegating *how* to achieve the results, just talk about the results themselves and encourage initiative.

- Don't limit the concept of delegating to work chores; consider it also for the multitude of home chores.

- Don't do it if you can delegate it. Just recognize that you are delegating not only work but responsibility for results.

- When delegating, provide as much information about the task as possible.

- When you delegate, be sure you delegate the authority to make important decisions.

- Tell the truth to the person to whom you delegate. If it is drudgery, don't say the task is glamorous.

- If you don't know how to trust, you'll have problems delegating. So be sure you learn how to trust, how to give up some territory and power.

Of all the tools necessary to enjoy a wealth of ninety-minute hours, delegation is the toughest to master. However, I guarantee that it is worth the time to hone your skill at.

5

Using Audiotapes to Gain Extra Time

Most authorities on economics would agree that the more you learn, the more you earn. Study after study proves that college graduates earn considerably more than, usually double, what high school grads earn. Small wonder that millions of earthlings are welcoming the Information Age as it sweeps across the globe—the whole world, thanks to satellite transmission—like a tidal wave of knowledge.

What do you suppose is the number-one hobby among top-level Fortune 500 executives? The answer is reading. Some of it, no doubt, is for escape. But the majority of that reading is devoted to learning. These executives don't necessarily learn to earn, but it's likely that their earnings will rise on a scale that accurately reflects their knowledge.

The simplest, most obvious, and most available of the new timesaving technologies to begin using is audiocassettes. Around 90 percent of Americans own audiocassette players, with a healthy chunk of those existing as automobile units and personal headphone cassette players—the two versions of cassette players most appropriate to the purpose of time extension. Cassette players can be bought for under twenty dollars. The combination of the burgeoning audiocassette market and the omnipresence of cassette players gives everyone ready access to the ninety-minute hour.

SPEED LISTENING DOUBLES YOUR LEARNING

In fact, by the use of cassettes plus a device that speeds up the sound track without making the announcer sound like a chipmunk, you can get up to 180-minute hours!

Here's how that arithmetic works:

- You drive home from work, say a sixty-minute drive.

  TOTAL REAL TIME: 60 minutes

  TOTAL TIME VALUE: 60 minutes (Drive only)

- While driving, you hear a sixty-minute teaching cassette.

  TOTAL REAL TIME: 60 minutes

  TOTAL TIME VALUE: 120 minutes (Drive plus cassette)

- You listen to the cassette at double speed using a time-compression machine. When it's over after thirty minutes, you put on another sixty-minute cassette and hear that at double speed.

  TOTAL REAL TIME: 60 minutes

  TOTAL TIME VALUE: 180 minutes (Drive plus two cassettes)

That's a whole lot more than a mere ninety-minute hour, yet it illustrates the point of doing one thing while doing another. It also illustrates how new technology exists that enables you to do many things at faster speeds than before.

If you want something enough, you can *will* it to happen. And if you can will it, you can take actions to *make* it happen. One of the most effective courses of action is gaining information, eliminating ignorance, picking up new truths, thereby obtaining an enormous advantage over competitors who lack this information and insight.

THE IGNORANCE OF THE FIFTY-YEAR-OLD

Many educators claim, with due justification, that the average college graduate is better informed than the average fifty-year-old. This is true because the fifty-year-old's formal education probably ceased when he or she was about twenty-one. Since that time, a whopping twenty-nine years, a great deal of new information has been unearthed, discovered, and created. Unless the fifty-year-old regularly reads every periodical on every subject, there's a good chance that he or she misses out on a lot of valuable information. But the cream of that data gets spoon-fed to the under-twenty-one generation in the form of formal education. In school, they learn Basic Living 101 stuff that you've never even heard of.

It's hard to keep up. It's almost impossible to learn the most significant findings in every subject, but you can at least give it a try.

I've always felt that the minimum daily requirements for trying include reading at least one daily newspaper, watching at least one daily TV news show or listening to at least one daily radio news show, reading one weekly newsmagazine, and reading at least one weekly or monthly magazine in your own field of expertise. That might not make you as well-informed as a twenty-one-year-old, but it would furnish you with a body of knowledge sufficient to be well-informed.

To supplement that meager education and to fuel your ambitions on the way to your objectives—not necessarily your day's objectives, but your life's objectives—it makes a lot of sense to gain the information available on audiocassettes. The topics they cover range from business to psychology to sports to daily living to sex to—well, I've put a generous list of topics covered in the Appendix, and you'll see for yourself the wealth that awaits your ears.

WHERE TO START LIVING NINETY-MINUTE HOURS WITH AUDIOTAPES

First, I want to open your mind to the opportunities you have to gain this information, and to the many places where you can take advantage of them. If you don't already, you ought to own a car cassette player, a portable cassette player, or a personal headphone cassette player (but don't use this while driving).

Let's start with ten opportunities readily available to you to obtain information on audiocassettes:

1. While driving to or from work, to or from anywhere
2. While taking a bath (not a shower, but a bath)
3. While bicycling
4. While jogging
5. While walking
6. While doing many other forms of exercise
7. While shaving or putting on your makeup
8. While flying, on business or pleasure
9. While waiting for virtually anything virtually anywhere
10. While commuting by train

Here are ten more opportunities:

1. While doing the dishes
2. While doing other chores around the house
3. While engaging in certain hobbies: model-plane building, stamp sorting, making stained glass compositions, doing needlepoint, painting, sculpting, and lots more such activities
4. While commuting by car pool
5. While relaxing
6. While in the bathroom
7. While in a state of altered consciousness, such as self-hypnosis—about which more will be said in Chapter Eleven
8. While working in the garden

9. While cooking
10. Instead of watching television

It's not hard to see that there are numerous opportunities for 90-minute hours. And in the twenty cases just mentioned, I'm talking of 120-minute hours, since you'll be able to accomplish 60 minutes of whatever you're doing while gaining the 60 minutes of information put forth on the tapes.

You can use the information you gain in any way you wish: business advancement, increased earnings, professional skills, sports expertise, personal enrichment, improved relationships, problem solving. When you look over the list of information available on audiocassettes, I'm sure you'll spot many opportunities to improve your life. And if you combine this learning with activities that do not require your full concentration, you will be gaining knowledge without giving up time. You will be extending your hours to even more than ninety minutes.

LET'S HEAR IT FOR TIME-COMPRESSION MACHINES

First, you should be aware of a new device currently priced by some companies at around one hundred dollars and available at up-to-date electronics retailers as well as through several mail-order catalogs. I'm reluctant to mention names because by the time you see this page, the machine may cost less than twenty-five dollars, the way electronics prices are dropping. And it may be available at your local 7-Eleven.

I'm referring to a time-compression machine. It enables you to play your audiocassettes while electronically removing the spaces between the words spoken, so you hear more words in less time. With the variable rate control on the machine, you can increase the speed of delivery from 1 percent faster to 100 percent faster.

The announcer's voice remains at the same pitch. Only the speed of delivery changes. Now I know that you've heard sound tracks speed up until the deep-voiced announcer sounds like a gelded chipmunk, but I assure you, technology has solved that problem. Not only do you hear the exact words at the exact pitch of their original delivery, but because the pauses are electronically eliminated, your retention of the words actually increases.

Columbia University did a study in the late 1970s which proved that people who hear fast-talking announcers, speaking at rates of over 200 words per minute (the average speaker delivers words at a rate of about 140 per minute), tend to retain the ideas communicated far longer and in greater detail than those who hear smooth but slow-talking announcers. We may not like fast talkers, but we listen to and remember them. Slow talking is just too languid for the human mind. Even 280 words per minute is not too fast for our minds.

This all makes it plain to see that a time-compression machine enables you to gain twice as much information as a standard cassette player. Naturally, you won't be able to use one if you planned to jog wearing your personal headset. But you can use one in your car, at home, or on an airplane. Don't worry about the other passengers staring at you as you gobble up information at world-class speed. You can always wear headphones. If your cassette sounds would be heard by anyone else, I highly recommend the use of headphones (although, again, not while driving). After all, you don't want the whole world to learn about effective negotiating at the same time you are.

LISTEN TO WHATEVER YOU WANT OR NEED TO HEAR

Before I go into eye-opening details about the wide range of subjects available to you on audiocassette—areas in which you can easily obtain growth via information without taking up your valuable time—you should know that if you don't see the exact tape you want, you can make it yourself. Or you can go to one of the companies that will make the tape for you. They'll do it with standard tapes and with subliminal tapes—an advanced method of learning that will be examined in the next chapter.

One of the most successful companies I've ever witnessed in action makes audiocassettes of their top salespeople delivering an actual sales pitch. Then they distribute the cassettes to their huge sales force, suggesting that the salespeople listen to the tapes over and over while driving between clients. This gives the company and their sales force the benefit of real-life examples by proven successes, economy in the production of valuable cassettes, increased sales—plus an ample supply of ninety-minute hours.

This same company uses audiotapes to help their salespeople critique themselves. But that's merely an item of interest and not a tidbit to put more minutes into your hours. It is, however, a reminder that more and more growing companies are relying on audiocassette technology to inform and prosper.

NEVER WRITE ANOTHER MEMO

Audiocassettes and players do give you access to ninety-minute hours in other ways besides instructing you in state-of-the-art living. For anyone who writes memos or letters, they are also handy dictating machines. Why wait till you get to the office to write or dictate your memo to a secretary? That's what Stan Danton, antihero of Chapter One, would do. But we know what a timewaster Stan is.

Simply dictate your memo or letter into your cassette player, which should also be a recorder but probably isn't if it's a personal headphone stereo. So be sure you either own a portable player-recorder or have one installed in your car. It can pay for itself many times over. Then you can hand your business correspondence to your secretary, who will transcribe it into finished form. This requires no extra time on your part because you accomplished it while driving. Such are the advantages of ninety-minute-hour living.

The technology already exists—but alas, not in the price range of the everyday American business—to insert a dictation cassette into a computer with voice recognition capabilities. The computer prints out the finished piece or pieces, and even the secretary gets in on the benefits of the ninety-minute hour.

Tiny microcassette player-recorders, of which I write in Chapter Ten, also enable you to enjoy the timesaving luxury of ninety-minute hours. But I am keeping them out of this chapter since most audiocassettes are not yet available in this teensy format.

One of the great benefits of working on my own, rather than in the huge corporate structure, is that I am protected from the time demands of memos. I might feel different if, during my corporate days, I could have dictated my memos on the way to work and listened to the memos of others on the way home from work. That's the kind of streamlining that will achieve some of the cost-cutting goals of the conglomerates, since executives waste so much time writing and reading memoranda.

The ninety-minute hour certainly works in the direction of cost cutting by the more efficient use of time.

THE NINETY-MINUTE-HOUR COMMUTE

Right now, even though millions of American workers commute by train or bus, zillions more drive. I know that for a fact because I've seen at least one zillion, or maybe it was only a trillion, in the lane in front of me practicing slow driving and testing their brakes.

The vast majority of solo motoring commuters listen to the radio, enjoying music, news, or a talk show. I am all for music while driving, but I am even more for music mixed in with a few instructional audiocassettes during the week. Similarly, I am all for news, but I'd just as soon get it during the late-night TV news show or from the morning newspaper if I could become smarter simply by playing an educational audiocassette instead of listening to the radio news show in the car.

If you don't drive during your commute, you can practice trainbound ninety-minute hours by reading business matter or books, or by listening on your personal headphone to audiocassettes. The idea, again, is to combine activities to give yourself extra time. It makes no sense to use your commuting time simply for riding back and forth like a mannequin. You can double the usefulness of that time by reading, writing, dictating, or listening to an outstanding variety of audiotapes. You can elect to entertain or enlighten yourself. And it won't take one extra second of your lifetime. You can transform all those 60-minute hours into 120-minute hours. That is almost like adding extra hours to your life.

A LIBRARY FOR YOUR EARS

Audiocassettes are available at bookstores, libraries, department stores, drugstores, convenience stores; through catalogs; via direct-mail offers; and even in many vending machines.

Let's look at some specific audiocassette titles.

I'll start with languages. Just one company, AMR, which stands for Advanced Memory Research, has been building a sterling language-by-tape reputation over the years. They offer your choice ($125 for eight 45-minute cassettes at this writing) of Spanish I, Spanish II, French I, French II, German I, German II, Italian I, Japanese I, and Chinese I. The company claims they can give you a working knowledge of a foreign language while you drive to work, while you jog, or in the evening before bed. An enticing feature of this and all cassette learning: You set the pace.

Because of the teaching methods they use—and you'll experience several exciting methods—many cassette producers explain that no reading or rote memorization is required. Unlike other language teachers, they tell you that you don't even have to pay close attention. As one who has lost a fair amount of weight through the magic of the ninety-minute hour and audiotapes, I can personally attest to the truth of that statement. I paid no attention whatsoever to my weight-loss tape, yet the pounds came off with incredible ease. I was, I admit, listening to an audiotape peppered with subliminal suggestions, the topic of the next chapter. Still, I liked the idea of not having to concentrate on my goal. That was contrary to everything I had heard about learning in school.

The AMR people say that their foreign-language cassettes will teach you twelve hundred words—twice the number of words used in ordinary conversation—along with the basics of grammar. To prevent boredom, they furnish you with a minimum of six speakers from different regions where the language is spoken. And they add music, since music relaxes you and promotes fast learning.

One of the most winning features of ninety-minute hours with these audiocassettes is that you can learn the language you wish to learn in your spare time. That's the part that gets my attention, that stirs my imagination, that ought to stir yours, and that leads one to believe that the world is irrevocably headed into a universe of ninety-minute hours.

The range of nonfiction audiocassettes available offers as much information as a college education. They can provide you with the data to accomplish the stuff of your dreams. And they ask for none of your free time in return for their words of wisdom. You'll be pleased to know that the gamut of fiction, from whodunit to romance, Mark Twain to spy thriller, is also available on audiocassette. If you begin to fill your life with ninety-minute hours, you will have the time to enjoy these too.

6

Unleashing the Power of Subliminal Suggestion

You don't have to lose out on the contributions that audio-tapes can make to the ninety-minute hour just because you are not a lone commuter with time to listen to tapes. Many people will appreciate audiocassettes strictly because they offer access to the honestly awesome power of subliminal suggestion. They will make this power the most valuable contributor to their ninety-minute hours. This is because

- It requires zero concentration.
- You are actually encouraged to do something else while receiving subliminal suggestions.

- It is an extremely stress-free route to the ninety-minute hour.

- It is truly pleasurable.

- It works with amazing effectiveness—if you want it to work. But it won't work if you don't want it to.

By combining the common audiotape learning technique, conscious listening, with a decidedly uncommon psychological learning technique, subliminal suggestion, you can begin to move at hyperspeed in pursuit of one set of goals while you also zero in on others. Pursuing two goals at the same time shifts you into a pure ninety-minute-hour mode. And the subliminal method of saving your valuable time provides you with as much relaxation as it does enlightenment.

By blending simple hearing with subliminal stimulation, you are empowered with the remarkably impressive force of your unconscious mind. If you're not yet aware of its force, you've got a happy surprise in store for you. This is a bonus of the ninety-minute hour. Along with its time-saving value, subliminal suggestion provides you with a potent energy source you may never have tapped before. Most people haven't.

IT'S TIME YOU KNEW ABOUT SUBLIMINAL SUGGESTION

Subliminal messages are directed beneath the threshold of your consciousness. They are aimed with precision at your unconscious mind. The word *subliminal* itself comes from two Latin roots: *sub,* which means below or under, and *limen,* which means limit or line or threshold. Anthropologist Victor Turner in *Dramas, Fields, and Metaphors* focuses on "liminal" or threshold states. He suggests that in stable societies thresholds, or "limina," may be rites of passage, initiations.

On audiocassettes subliminal messages are implanted, really "hidden" from your conscious perception, within a background sound that can range from music to waves, from winds to the words of your favorite TV show. The messages are placed at a slightly lower decibel level than the background sounds, so that they can "sneak" undetected right into your unconsciousness. But just because you aren't aware of hearing the messages doesn't mean you don't "know" them. You do. Subliminal production technology has enabled tape makers to synchronize and match the volume of subliminal messages with the volume of the background sound or music.

One of the keys to the success of subliminal messages is their continuous repetition. Hearing such simple messages consciously might be enough to drive a person up the wall. But "hearing" them subliminally is a different matter altogether. It is a well-known fact that repetition is one of the most efficient manners of accessing the unconscious mind—yours or anyone's.

Technology has enabled us to intentionally cross the subliminal threshold since the early 1950s, but social and legislative obstacles have been erected because of the guises, invisible to the public, that subliminal suggestion could take. Consider this ability to penetrate the mind of an unknowing public in order to sell a soft drink, an automobile, a president of the United States.

Although Dr. O. Poetzle first discovered the concept of subliminal perception back in 1917, the real idea of subliminal stimulation became clear in the American consciousness in the mid-1950s in a study run by James Vicary. A New Jersey movie theater flashed the words "DRINK COCA-COLA" over Kim Novak's face during a six-week run of the film *Picnic*. The words were on the screen such a short time, a tiny fraction of a second, that the audience was not consciously aware of them. But Coke sales in the lobby increased 58 percent!

Naturally, this aroused the consumer watchdogs, and believe me, there weren't many of them in the midfifties. Still, the "Big Brother" possibilities of a tool such as subliminal suggestion motivated the Federal Communications Commission to institute a policy against "concealed" advertising. Even so, they passed no rule against it.

Contemporary companies offering audiotapes and videotapes with subliminal suggestion do so only with your consent. Nobody is trying to sell you anything. (Still, I wonder why I eat so many Hydrox cookies while listening to my Paul Simon *Graceland* album.)

The ability to influence people without their knowledge put subliminal suggestion in the same category as mesmerism and hypnotism. For too long, it has been kept in the communications closet.

Don't let subliminal power pass you by just because you can't hear or see it.

Subliminal learning power is emerging, and with it come new opportunities for you to learn, to achieve, to star, to earn, to solve problems, to master skills, to tap into those higher forces, that loftiest power that resides within you. One of the most winning aspects of subliminal suggestion as a timesaver is that the companies producing these tapes actually tell you to do something else while you're listening. Read this quotation from a current large producer of subliminal tapes: "Don't stop and listen to our tapes. When you get our tapes, play them while you're doing something else. Listen to them as you work, as you play, as you exercise or watch TV. Or don't listen to them at all—simply play them in the background as you do what you usually do during the day. They work whether you pay attention to them or not."

Because this statement goes on to address the issue of freedom from stress, I'll complete it: "Point is, programming yourself is as easy as we said earlier—and . . . it doesn't even take extra time to do so. It almost sounds too easy, doesn't it? You expect to have to stress yourself to change yourself, don't you? Well, this time, no stress needed. Fundamental change finally comes easy."

It all puts the ninety-minute hour more and more within your grasp. And it shows how little you have to lose in an attempt to reach for the brass ring, the one made out of extra time.

If the remarkable upswing in Coke sales in New Jersey while William Holden was romancing Kim Novak hasn't intrigued you, here are a few more substantiated facts about subliminal suggestion that ought to convince even the most hard-nosed skeptic:

- A department store, according to a story in *Time* magazine, implanted the subliminal suggestion "I take a great deal of pride in being honest. I will not steal" in its Muzak system. Over a nine-month period, shoplifting was reduced by 37 percent.

- *The Wall Street Journal* related a similar story of a marked decrease in shoplifting in a New Orleans supermarket. The subliminally implanted message was "If I steal, I will go to jail." Pilferage skidded down from $50,000 to $13,000 every six months. Beneficial side effect: Cashier shortages dropped to less than $10 weekly from $125.

- In *Preconscious Processing* by Dr. Norman Dixon, 748 references on subliminal learning are cited. Positive results are revealed in 80 percent of the cases. This proves that subliminal suggestion is not perfect, but it can accomplish startling results.

- In 1980 the McDonagh Medical Clinic in Gladstone, Missouri, completed a seven-month trial of subliminal messages put forth in the waiting room. The results: Fainting was reduced to nearly zero; smoking was reduced 70 percent; temper flare-ups went down by 60 percent.

- To test the efficacy of subliminal suggestion, I decided to try it for myself. I purchased a subliminal weight-control tape and did only two things daily:

1. I spent fifteen seconds, no more I swear to you, reading the subliminal suggestions aloud. (They were printed on a card that came with the tape, which set me back $9.95. It's now $12.95.)
2. I listened to the music each day, while reading, working, goofing off, anything but consciously listening. What happened? What happened is that I lost fifteen pounds within sixty days without one iota of willpower. I didn't know I was losing weight; I wasn't trying to lose weight. It just seemed that I was less hungry at mealtimes, less inclined to go for that bedtime snack. I certainly never went on a diet. But I did lose weight. So I can credit only the magic of subliminal suggestion for

my trim new bod. Thank you, psychologists, one and all.

Of all the tools for the ninety-minute hour, the power of subliminal suggestion seems the most advanced, exceeded only perhaps by the tool of your own unconscious mind.

THE 120-MINUTE HOUR AND SUBLIMINAL SUGGESTION

Subliminal suggestion takes absolutely no time at all, giving you the benefit of many 120-minute hours. This is time during which you devote effort toward your primary work while listening to music. If that causes stress, you're in deep trouble. If that's too cumbersome to get involved with (like so many "timesaver" calendar-datebook-diaries), you're destined to waste a great deal of the limited time in your life.

If I were you, I'd check into this technology very closely. In fact, I already did, and I promise, it's a winner—if you sincerely want to accomplish the goals set forth by the tape.

A funny but true aside: One of the subliminal catalogs warns people that if they don't want to lose weight they should not listen to the tape. If they do hear it, the catalog copy warns, they'll lose weight regardless of their desires. I quote: "WARNING: This subliminal tape program is a powerful tool. Do not repeatedly expose individuals of normal weight, especially children, to this tape program. Repeated exposure may cause unwanted weight loss in these individuals."

Subliminal images are totally invisible, even on videotape. To hammer this home, I need only remind you of the New Jersey movie house with its suggestions superimposed over film. Today's subliminal videotapes enable you to achieve specific goals subaudibly—with the message just out of your hearing range—and subvisually—with the message flashed to your brain without conscious recognition at a rate of one-thirtieth of a second.

STATE-OF-THE-SCIENCE SUBLIMINAL TECHNOLOGY

To bring you up to speed in the whole subliminal industry, let me fill you in on other developments worth your consideration, especially if you own a business or care like crazy about the success of a business:

- Some subliminal tape companies promise "two-level power learning." They let you hear the message audibly (level one), then you hear it subliminally (level two).

- Many companies let you select the background sounds over your subliminal suggestions: ocean waves, stereo music, and lots more.

- You can choose whether the subliminal messages are delivered (even though you can't hear them consciously) by a male or a female—or both.

- You need not hear a whole subliminal tape to derive the benefits it offers. You can hear the beginning, the middle, the end, or only a few minutes. The messages are being conveyed constantly.

• Subliminal tapes are now available not only for hard-driving businesspeople, and efficiency-minded nonbusinesspeople, but also for kids. The kid tapes are oriented to stopping substance abuse, instilling learning and study habits, building self-esteem, gaining a sense of responsibility, liking school, feeling loved, and reducing stress. And, of course, there's also potty training for toddlers, and one called *A Messy Child's Guide to Neatness.* Sounds a lot more sensible than spanking.

• Some new tapes feature a male voice track directed to the left brain, where logic and sequential reasoning reign supreme, along with a female voice track directed to the right brain, where mood and emotion rule the roost.

• Several subliminal audiotape producers will be pleased to customize a tape for you, utilizing the exact messages you want, combined with the sound of ocean waves or music. Some ask for a brief description of your desires, ambitions, or problems—or your company's desire to energize employees, promote courtesy, discourage drunk driving, achieve your choice of goals—from which they'll design a subliminal tape (time compressed if you want) for your exact needs. The price, around $230.00 to $250.00, seems high compared with $12.95 for a noncustomized tape. But look at it like this: You're paying for the technology rather than a mere audiocassette. And if you or your business can derive a substantial profit from a $250.00 investment, it could be the wisest investment you've ever made. After all, it's only money; it's not time you must invest. And it is time you'll be saving.

- Still other subliminal tape producers utilize what they term a "phase modulation procedure," whereby a speaker's voice is treated by spectral analysis, then broken into subharmonic chord components. Next, the subliminal voice is filtered, synchronized, and blended with music. The final result, say the producers, is a synthesized subliminal voice that becomes an integral part of the music. They also inform us that the unconscious understands this combination of music and nonlinear language best of all.

- Subliminal suggestion, featured in this chapter, and time compression, explained in the last chapter, make a powerful combination for stimulating and informing your unconscious mind. Time compression can give you the benefit of 180 minutes per hour: 60 minutes spent doing whatever you want to do and hearing a 120-minute time-compressed subliminal tape. That's a total of 180 minutes of accomplishment crammed into an hour. And that hour might be devoted to working, playing, hiking, relaxing, anything—while listening to soothing sounds or beautiful music. For a moment there, I thought of titling this book *The 180-Minute Hour.* But after giving the matter some thought, I decided that 90 minutes is enough. Just remember that 90 is a conservative figure.

- It is now possible to gain the advantages of subliminal suggestion while listening to the radio—or watching TV! Subliminal cassette players are available that allow programming using special tapes while you listen to a radio station or view a TV channel. It sounds complex, but it is really very simple. For instance, while you watch your favorite TV show, visual subliminal messages are flashed on the screen every five seconds. They appear for so short a time that you do not consciously see them, but take my word, your unconscious mind will see them and learn them. All you need is a videocassette player and a TV set. You get to watch TV while learning whatever you want to learn. Even the kids can develop a love of school and learning while they watch Saturday morning TV fare. They may think they're having fun, but you know they're improving their lives at the same time.

• Some subliminal tape producers now offer what they term "holographic sound"—three-dimensional sound moving in and around and above your head. The concept is interesting, but if I were playing a subliminal tape while working on my word processor, I think I'd be happy with plain old two-D sound.

• Now you can purchase what are called "Self-Talk for Success" tapes with a five-minute guided relaxation section followed by an opportunity for you to create your own mental image consistent with the goal of each tape. While you are doing your imaging, the speaker supports that image with self-talk statements in the background. The selected music is said to entrain both sides of the brain. The producer calls this "full-brain learning."

- Reaching into the most current findings in the psychology of learning, one subliminal audiotape producer has combined guided relaxation, visualization, affirmation (more about these last two in Chapter Eleven), and subliminal communication into one program—covering a host of topics. As long as these techniques have been proven effective and are available, why hold back? One company didn't. They are devoted to reaching both the conscious and unconscious portions of your mind. But I am interested in saving time for you while increasing your accomplishments, so let me caution you against using subliminal tapes that take your time rather than save it. I'm all for learning, but not at the expense of your time. Subliminal learning is dynamite because it can work its wonders while you are doing your work. Adding too many other elements are what my grandfather used to say is "like putting raisins in the matzos."

By the time you read this, I wouldn't be surprised if ten new subliminal technologies have hit the market. But don't miss the main point—that the biggest plus of subliminal technology, when it comes to the ninety-minute hour, is its ability to do its thing while you do yours.

LISTENING TO THE SOUNDS YOU LOVE

In addition to teaching you things painlessly and with zero drain on your time, subliminal tapes allow you to hear exactly what you want to hear—except for total silence. Just consider the background sound selections now available to you:

Gentle winds	Inner peace
Easy listening	The Top 15 of 1750
Romantic moods	Lullaby music
Classical music	Synthesizer music
Steve Halpern music	Alive and well—Rock background
Healing music	Jazz
Largos and adagios	Astral sounds
Crystal fantasy	New Age music
Contemporary rhythms	Harp and flute
Tropical ocean	Reggae
Thunderstorms	Bluegrass country

These are just some of the background sounds and music available. There're more if you want more.

SAY SOMETHING IN SUBLIMINAL

The language of subliminal learning is primer simple. For instance, a stop smoking tape might contain these statements:

Smoking does not interest me.
Smoking stinks.
I break the habit.
I can do it.
Smoke tastes bad.
I am proud of myself.

Let's examine the words for relieving back and neck pain:

I am healthy.
Neck relaxes.
Back relaxes.
Pain is gone.
My body is powerful.

Just imagine the powerful effect of hundreds upon hundreds of repetitions of those statements, subliminally, to a person with constant back and neck pain. That person is hearing soft, relaxing music while, say, working on next year's departmental budget. Imagine the happiness of the listener when eventually his or her back and neck pain disappears. It took no time, yet exerted a positive influence on the listener's life.

What do you suppose a time-management subliminal tape might offer up as clandestine statements?

I create quiet time free from interruptions.
I am master of my schedule.
I find what is important and give it priority.
I recognize and deal with conflicting time demands.

This particular tape, for instance, is available with background sounds of harp and flute, easy listening, waves, or free and clear just waiting for you to plug it into your VCR. The cost is $10.95, plus two bucks for postage and handling.

WHAT CAN YOU LEARN SUBLIMINALLY?

Asking what you can learn subliminally is almost like asking, "What kinds of books can be found in the library?" These days, subliminal offerings come in a wide variety of topics, from business to health, personal growth and relationships to sports and improving sex. To give you an inkling of what the ninety-minute hour can mean to you, spend a fraction of that time considering the many subliminal possibilities listed in the Appendix. Go ahead, do it now; I'll just wait here.

Pretty impressive, eh? As you can see, comprehensiveness is not a weak spot in the subliminal offerings of today. If you're getting the idea that subliminal suggestion is becoming a big industry because it works, you're getting the right idea. And if you're realizing that subliminal suggestion can help you without using your ultravaluable time, you are right again.

Consider the possibilities of a new and inexpensive technology that can merge up to six and even more inaudible suggestions with audible music or sound effects—all with the goal of making you a more successful human being. Consider how subliminal videotapes can communicate invisible messages within the context of visible visual entertainment. Here you are watching "Wheel of Fortune" while learning all about *Prosperity/Living the Dream.*

A final recommendation other than to give this brave new technology a try in your quest for ninety-minute hours: Use the tapes one at a time. *Human Potential* magazine, in its January 1986 issue, ran an article stating that subliminal tapes addressing addictive behaviors, such as overeating, smoking, drinking, or drug abuse, should not be used at the same time because the body may unconsciously resist sudden pressures to break several habits at once. The article counseled readers that subliminal programs are more effective when used one at a time over a period of several months.

So here you are, learning how to break your cigarette

habit and improve your golf game when all you really wanted was a way to pick up a few extra minutes a day. Now that we've changed your life, let's go back to saving your time.

7

Using TV to Put More Time in Your Life

It doesn't seem to make sense to include a chapter on watching television in a book about creating extra time. Where are the timesaving opportunities with television?

They're right there on your screen once you take a moment to stop thinking of TV as a forum for "Dallas" and little else. Television has changed dramatically since you were young. And it's going to continue to change dramatically. As television emerges, it is beginning to offer new opportunities for entertainment, instruction, uplift, purchasing, insight into current events—and for saving you oodles of time. And that's just the modest start.

But just as you cannot practice ninety-minute-hour

principles without people to whom you can delegate your work, neither can you begin to add extra minutes to your life by means of audiocassettes and subliminal stimulation if you don't own audio- and videocassette players. To get the most that TV offers, you'll have to incorporate modern TV equipment into your life. You may have traded in that old, tiny black-and-white set years ago. But viewing in color, though pleasurable, is hardly a timesaver and certainly not the fastest way to the ninety-minute hour.

STATE-OF-THE-MOMENT TELEVISION

Chances are you're going to become fascinated with the possibilities now offered by the latest developments in television. But before you rush out and invest in them, be warned that the medium is changing so fast that the new developments of today will probably be replaced by newer developments tomorrow. One quantum leap not likely to be surpassed soon was the advent of satellite TV, which is not yet in the mind, much less in the homes of the majority of the American public.

As you can surmise, the ripening of television technology will be a continuing process, so I suggest that if you want to get the most that television offers, you'll have to bring yours up to date. That doesn't necessarily mean that you've got to have the biggest or the brightest picture. But it does mean that you'd be wise, in pursuit of the ninety-minute hour, to consider adding to your TV arsenal such advancements as

- A videocassette recorder
- An earth station
- Digital television

- A video camera
- Instructional videotapes

Gaining access to the ninety-minute hour, sometimes inexpensive as with subliminal tapes, sometimes more costly as with television, generally requires that you supply equipment and commitment. In this case, they turn your TV set into a time-creating machine.

Take a few seconds now to think through the equipment side of that equation. The more TV enhancers you own, the more time you can save. Let's take a closer look at the five new TV enhancements I just listed.

YOUR VIDEOCASSETTE RECORDER

With well over half of the households in the United States owning VCRs these days, and sales climbing at a rate of one million new VCRs per week, there's a fifty-fifty chance that you've got a VCR already, maybe even two. To save substantial portions of time, your VCR ought to be outfitted, as most are, with programmability and a remote control. You might also want to have a new device, which cuts VHS tape rewind time in half and also allows you to view one tape while rewinding another. Devices like this cost under fifty dollars while saving and granting you access to many minutes.

Videocassette recorders save time for you by putting you in charge of programming TV fare to your tastes and time availability. You've probably got a lot to do without having to live the Gospel of Life According to *TV Guide*. You can watch what you want when you want, and you can skip past the parts you don't want to see. I quote from a column in today's newspaper. The column refers to the final fifty-eight seconds of a National Basketball Association Championship game: "Later, I replayed the final 58 seconds on the tape machine. Total elapsed time was 14 minutes. That included 11 commercials." Had the columnist taped the game to watch later, the final fifty-eight seconds would have taken fifty-eight seconds and no more.

By calling the shots on television programming, you can adjust the medium to fit your mode. Using television this way, you'll soon discover its ability to fill in spare time rather than require its own part of your day.

To give a minor example: if you want to watch "Cosby," which is on at 8:00, and you're through with dinner at 7:45, you've got to kill time until 8:00. Time is an endangered species; it is not meant to be killed. If you tape "Cosby" to watch the next day, you can start it when *you're* ready, not when *it's* ready. That takes only fifteen minutes, but when you multiply those fifteen minutes by the number of TV shows per year for which you kill ten or fifteen minutes, you can see how this one small application of a VCR can create more time for you. You lose six and a half hours annually just for "Cosby"—and that's not even counting reruns. Almost all television programming (except the news) is not that timely and can be moved to times better suited to your needs.

You can also easily eliminate parts of shows, such as halftime during sports telecasts. Once you become adept at programming your VCR, you'll become privy to the second major timesaver of owning a VCR—zapping.

Saving Time with Zapping

When you are viewing a videotape and on comes a commercial, using your remote control to fast-forward through the commercial is known as zapping. As an ex-TV commercial writer with a certain respect for television ads, I feel of two minds suggesting that you do your very best to eliminate TV commercials from your life. Still, I call on you to zap mercilessly, using this efficiency to create extra time for your life. I feel a lot more respect for your time than I do for television commercials. These days, during prime-time television, six minutes of every thirty are devoted to commercials or station promotional material. These are easily zapped.

You can create, via zapping, even more time when you tape material telecast during off-time viewing hours—before 7:00 P.M. and after 11:00 P.M. Frequently, you can happily zap away fifteen minutes of each hour. Since the trend is toward allowing even more commercial time per hour, this gives you even more zapping opportunities, saving you even more time.

The average American family keeps a TV set on more than sixty hours a week—nearly nine hours a day. So if you're an average-to-heavy TV watcher, we're talking about saving megatime each year simply by zapping. What might you do with those extra hours? Not watch more TV, I hope. But you can. Free time is just what it says it is.

It takes about three hours and fifteen minutes to telecast most pro football games. Zap out the time-outs, commercials, and half-time hoopla and it takes one hour. Zap out the periods of nonaction, when players are in the huddle or lining up or the quarterback is calling signals, and you can see an entire game in 10 minutes! That works out to creating an extra two hours and five minutes of time to be spent any way you like simply by zapping.

Baseball is a leisurely type of game, with a graceful luxury in the slow pace between the furious blasts of action. But just in case you want to zap your way through one, you can figure on the three-hour telecast running about 15 minutes of heart-stopping action. If you want to get into the action only, zapping through baseball games on TV can save about two hours and forty-five minutes of your living time.

An all-out sports buff with a taste for ninety-minute hours can watch ten games in the same time it would take another sports buff, without ninety-minute hour-savvy, to watch one measly game.

Videocassette recorders help you speed up your viewing by helping you zap right on through the boring parts of shows as well. For instance, if you're a devotee of public television, you may be immune to commercials, but you're not immune to pledge breaks. Your handy remote will help you remove those interruptions in your life.

At their worst, VCRs can save 20 percent of your time—and that's just during prime time. During sports or off time they can save a lot more—frequently enough to add thirty minutes to each of your hours.

YOUR EARTH STATION

There are currently about 2 million satellite-dish earth stations in the United States, but that number rises dramatically on a daily basis. The cost, on an installment basis, of an earth station—which includes no more than a satellite dish, satellite receiver, descrambler, and TV set—is about the same as the cost of subscribing to a cable TV system. Plus you'll own the earth station after it's paid off.

The big difference is that your cable system may give

you access to up to twenty-five channels. An earth station enables you to view more than 150 channels.

Earth stations give you the unique opportunity to rearrange time, just as simply as VCRs allow you to rearrange programming. Because shows are telecast in different time zones, you can use your dish to aim at the satellite that is telecasting from the time zone most convenient to your needs. From the standpoint of TV, you are living in four time zones simultaneously. You can be the first on your block to see reruns.

Earlier, I said that the world is moving into the formation of two classes: the information class and the noninformation class. Needless to say, the information class will gain a multitude of benefits and privileges because of their information. You may not know what is available to you, free of charge, from outer space. Every hour you are offered, via satellite, programming that is the equivalent of a seminar, a course, a vacation, a lifetime full of learning and being entertained.

Videocassette recorders now hook up with satellite tuners so that you may program your VCR to record any show at any time from any satellite. There are twenty-seven communications satellites orbiting earth in the Clarke Belt, 22,300 miles up in space. There will be more by the time you read this. Each of those satellites is capable of transmitting on twenty-four television channels, called transponders in outer-space lingo. Some satellites are using all twenty-four of their transponders; some are using just a handful. Regularly, the whole gang of satellites collectively transmit on an increasing number of transponders.

If all the satellites in use today transmitted TV shows on all their transponders, you'd have a selection of 648 channels. Even if half were scrambled, the menu of TV fare would be bountiful indeed.

Earth station owners save time because they don't have to go to the video rental store. The movies they want are transmitted free by satellite. They save time because they don't have to attend seminars or take courses on the topics that are offered in their *Satellite TV Week*—the *TV Guide* of the satellite industry. (If you're interested in these offerings, call *Satellite TV Week* at 800-358-9997 to see if they'll part with a back issue. If you live in California, call 800-556-8787.) One reading of *Satellite TV Week,* which is also available at most well-stocked magazine stores, is enough to show you instantly the startling difference between standard TV and satellite TV. Dish owners tend to spend about half an hour each week reading though their guide to programming, then pinpointing the shows they don't want to miss—either by live transmission or via videotape.

I haven't even mentioned the enormous selection of home shopping shows now available not only on satellite TV but also on cable. These enable viewers to save money and loads of time by purchasing by phone after the TV sales pitch.

Teleconferencing, the ability of one presentation to be seen in multiple locations around the world, is another major satellite technology timesaving capability, especially for businesses. If you run a business and want to have an interactive meeting with associates or prospects throughout the nation, teleconferencing makes it relatively inexpensive, very easy to set up, and extremely time conserving. During the next few years, it is predicted that teleconferencing will grow at a rate of 30 percent. It is also expected to eliminate $800 million in nonproductive executive travel time by the early 1990s.

Hardly in the same ballpark as an earth station or teleconferencing, but still utilizing satellite technology to save your time, is a hand-held monitor called a Quotrek, which uses satellite transmission to give you stock quotes. For about $400 you can buy one of these and gain instant investment information on 7,000 stocks on the New York Stock Exchange, the American Stock Exchange, and the National Association of Securities Dealers Automated Quotations; futures on the Chicago Board of Trade, the Chicago Mercantile Exchange, and the Commodity Exchange Center, and options from the Options Price Reporting Authority. A Quotrek is a timesaving gizmo for any time-loving investor. Unfortunately, at present the Quotrek only works in Los Angeles, San Francisco, Chicago, New York, Dallas, Houston, Miami, Atlanta, Philadelphia, Boston, Phoenix, Washington, DC, and Baltimore. But don't forget—we're only in the infancy of the satellite TV industry.

YOUR DIGITAL TELEVISION

Digital TV enables you to view two shows at the same time. It's not hard to see how this can mean 120-minute hours for you and anyone viewing with you. Simultaneous viewing not only lets you view two shows at the same time but also eliminates the need to tape some shows, which you can watch while you're watching others.

It's a bit disorienting for some people to view two shows at the same time. Takes some talent and/or experience. But once you get the hang of it, you'll be adding extra minutes to your hours regularly.

Like the satellite dish owners, with a digital television

set you'll be gaining a far clearer picture than you enjoyed with old-fashioned TV. But as a timesaver extraordinaire, you'll be enchanted by the timesaving capabilities of digital TV. For under five hundred dollars, you can buy a digital stereo TV converter now and start enjoying twice as much TV without devoting twice the TV viewing time.

YOUR VIDEO CAMERA

Although a video camera can be used as an entertainment device, it offers much to you in the way of timesaving advantages. A videotape of our belongings for insurance purposes saved us a ton of time over the painstaking verbal descriptions most people have to supply. Because there is no time necessary to process the finished product, you have instant access to your results, and you can re-shoot if you're not satisfied.

In making movies, a shot is done on film. The film is then processed, and the next day the processed film is shown as "the dailies" or "the rushes." That's a twenty-four-hour wait to learn that you messed up somewhere along the line and have to do the whole thing again. As a result, video cameras are now present at filmings. They tape what the film camera films. Only instead of requiring a full day to learn the results, the video playback shows them instantly.

If you shoot home movies or slides, a video camera can prove extremely valuable because of its speed and simplicity. Its implications in business are even more widespread. Models of new designs can be videotaped and analyzed by computer to save time and money. And, finally, your video camera makes a far more convenient record of your trip, from the standpoint of both saving time in processing and saving time in showing the trip to your friends. Instead of setting up the movie screen and the slide projector, simply slip the cassette into the VCR, turn on the TV, and voilà! You're in France!

Video cameras, which record sound as easily as visual images, are also timesaving as an instructional device. If a friend aims your camera at you while you take a swing with your golf club or tennis racket, the video playback will reveal more horrible truths than any well-meaning pro. Chances are, the playback will show you how to correct your imperfections, may they be few in number.

If you're in a line of work that requires you to make presentations, give speeches, or deliver sales talks, a video camera can be your best friend when it comes to honest feedback. People who love you may miss some of your flaws; cameras catch all of them. Or in your case, maybe only both of them.

YOUR INSTRUCTIONAL VIDEOTAPES

These aren't the videotapes that instruct you as do the audiotapes I discussed in Chapter Five. And although I do recommend the use of subliminal videotape technology, especially the kind that gets invisibly superimposed over the TV shows you are watching, right now I am speaking of another kind of videotape instruction that dramatically speeds your learning.

Your learning of what? Of business skills. Of sports. Of self-improvement. One company that excels in this type of teaching is called SyberVision. I've had several personal experiences with SyberVision, including watching a friend view a SyberVision golf videotape just once, then instantly knock nineteen strokes from his golf game . . . at my expense.

Although SyberVision is concerned with performance, I am impressed with the *speed* at which it teaches you to perform. That's why I unhesitatingly recommend SyberVision videotapes as passports to the ninety-minute hour.

Let's let SyberVision's catalog speak for SyberVision:

The human nervous system has the unique ability to learn through visual images. Scientists have known for years that the learning of skills and behavior are primarily the result of visual learning. (You learned to talk and swim this way.) But how visual learning happened remained a mystery. Scientists knew that if they could discover the mechanism in the brain that converts images into skills, the results would be profound. And now the results—SyberVision, in a joint, six-year research effort with Stanford University, has discovered exactly how visual learning occurs. The brain follows the properties of a mathematical equation, called the Fourier Transform, which transforms images into skills and behavior.

SyberVision has applied this knowledge to a revolutionary new and easy-to-use system with . . . performance training right in your own home. With very little time and effort, you can dramatically improve your performance in your favorite sport.

And here's more:

Each week we receive hundreds of unsolicited letters from people who have had success with our products. In fact, independent research shows that nine out of ten SyberVision customers experience significant improvement after using our programs. . . . This year two hundred thousand skiers will receive SyberVision training at Vail.

Two hundred thousand? I hope they don't stand in front of me in the chair lift line.

Although SyberVision is justifiably proud of their audiocassettes entitled "Profiles of Achievement," "Achievement," "Self-Discipline," "Leaders," "Successful Marriage," and "Successful Parenting," and their six language programs (Spanish, French, German, Hebrew, Greek, and Russian)—which can all be played at double time with your time-compression machine or listened to while you are driving or flying—it is their video technology that is the real timesaver. You watch a sport being performed flawlessly by a professional. You see each movement in slow motion, over and over. The sound you hear is music. Your conscious mind sees the shots, but your unconscious mind remembers them and passes the information on to your muscles.

Many psychologists claim we have memory in our muscles. This unconscious message is what your muscles receive when you watch a golf tape, skiing tape, tennis tape, baseball tape, bowling tape, racquetball tape, "Men's Defend Yourself!" "Women's Defend Yourself!" or even "Weight Control and the Will to Change."

Sure, you might learn these talents with lessons. But lessons take time. Practice takes time. Travel to and from lessons and practice takes time. By contrast, it takes very little time to watch a videocassette that can give you a shortcut without shortchanging you.

One more personal experience I had with SyberVision proved its efficacy beyond any doubt. My wife had brain surgery for a large but benign tumor. Following the brain operation, she contracted pneumonia and had to spend thirty-three days in intensive care. When she returned home, unable to walk, she underwent daily physical therapy. Part of that therapy was teaching her to walk. To practice, she used our very steep, quarter-of-a-mile-long driveway. She hadn't been able to walk to the top of it for over ten years because of the damage done by the tumor. After three months of solid and intense physical therapy, she could make it up one-quarter of the driveway.

I decided to try incorporating the SyberVision technique into her therapy program. So I took the home movies (converted to videotape years earlier because I try to practice what I write) we had made during our more than thirty years of marriage to a video-editing studio. The video editor and I selected eight sections that showed my wife walking energetically, running, climbing stairs, even skiing. We edited them so that each repeated ten times, five at standard speed, five in slow motion. Then we put them all together, added appropriate and uplifting music ("Imagine" by John Lennon), and ended up with a twenty-minute tape. I showed it to my wife, saying nothing.

Later that day, she made it to the top of the driveway—for the first time since the surgery, the first time in over ten years! My wife, her physical therapist, and I now all believe fervently in the speed and efficiency of the SyberVision concept.

8

Extending Time
with a Computer

I'll start right off with the good news: This chapter is written for that 88 percent of the population who do not regularly use a personal computer.

I recognize that many (12 percent) of you are so entranced with your computer that you can think of five hundred ways a computer can save time, leading to a treasury of ninety-minute hours. I also recognize that one would have to spend a treasury of ninety-minute hours gaining the computer expertise you already demonstrate so deftly. This chapter is not for you, the computer masters. It is for the computer avoiders, the teeming generations brought up without a computer for a classmate.

Think of this as a pro-time chapter, not a pro-computer chapter. Nowhere in these pages will I suggest that you spend hours inputting your chicken curry recipe on a computer disk (which you'd have to spend more time to copy onto a backup disk), then learn to summon up the data on the disk and even print the recipe with eye-catching graphics when you are about to fix a meal. That is a pro-computer thought, even a pro-chicken curry thought. But it is not a pro-ninety-minute-hour thought.

Still, I have no choice but to admit that a computer is unquestionably a friend to the seeker of ninety-minute hours. As a buddy once asked, "How else can you make one million mistakes per second?"

Many onetime plodders are now whizzes because they learned how fast computers are. Naturally, that's what computers are all about: speed. Anything they do, they do faster than humans can. Almost any time you use a computer, you are moving into the realm of ninety-minute hours. Some computer applications move you up to 90-million-minute hours. But those applications are for the computer-friendly folks among us. These suggestions are for the computer apprehensive. We are in the majority.

FROM TECHNOPHOBIA TO NINETY-MINUTE HOURS

I hesitated to buy a home computer because I didn't want to spend the time learning how to use one, learning how to speak computer languages, and learning the details about computer equipment so I wouldn't make an expensive mistake. I refer to buying the wrong disk drive, keyboard, printer, modem, surge protector, hard disk, and floppy disks—basic equipment for people interested in the timesaving aspects of computers. And let's not forget the custom-made furniture to hold all this gear; it sure didn't fit onto my typewriter table.

Talking with friends and associates who owned computers, and occasionally reading articles or seeing something on TV, I learned why I ought to be considering a computer and what it might be doing for me. I compiled a list of questions.

Using the Yellow Pages to phone various stores, I asked my compiled questions, saved eons of time, and found myself in just the right place. There, I tried out several computers. I never asked the salesperson to recommend a computer, because that's too much like asking someone to recommend a religion. Instead, I told him my needs and he steered me to a choice that seemed sensible for those needs. He was right. I ended up with an Epson QX-10, because it was by far the simplest to use—I didn't even have to learn a computer language or take any computer instruction. Since that time, they say Epson has improved their hardware, if that's possible.

Insofar as computer terminology goes, unless you have a need to learn the names of your digestive enzymes to enjoy a hearty meal, you won't have to learn computer terms. Although you can save impressive amounts of time—in the areas of gaining information, processing words, handling mathematical problems, keeping records, publishing, or communicating—it is very possible to live a normal life with a fair share of ninety-minute hours and never set one finger on a computer keyboard or mouse in your life. Lest you think you must own a computer, remember well the words of Georges Pompidou: "There are three roads to ruin: women, gambling, and technicians. The most pleasant is with women, the quickest is with gambling, but the surest is with technicians."

Nobody understands as much as I do your need to gain the benefits of ninety-minute hours. But nobody wants less than I do for you to invest in something that will take far more time than it will save, then end up in your closet anyway.

I would urge you to invest in a computer if you can save identifiably large amounts of time by using it for gaining data, word processing, number crunching, record keeping, desktop publishing, or electronic communication. For instance, salespeople are said to gain 15 percent more selling time when they use computers for record keeping.

If you need a personal computer for any of these purposes and you were waiting for the prices to come down, as surely they have, I'd suggest buying now, with the confidence that you'll earn more than you'd save by further waiting.

Although computers and programs may become annoyingly slow to you after you're familiar with their use, I suggest going through that familiarity process before upgrading, if you ever do have to upgrade. As a born technophobe with zero computer experience, not even having pitted myself against Pac-Man on an airport computer game, I'm glad I went through the process. My original software, which seemed lightning fast at the store, appeared to me to work like a turtle after one year. It was easy, then, to upgrade to a faster software. But had I started with the more rapid stuff, I might have been too intimidated to get on such good terms with my computer.

If you're like most Americans, you have this gnawing feeling that you are destined to own a computer and that it can save you time even if you don't do any of the six tasks I keep mentioning. I feel compelled to call to your attention the guy who suggested that the lasting benefits of the computer age on our society may ultimately prove as great as those of the CB radio. And keep in mind author John Bear's ninety-minute-hour observation: "Many people use computers to do things far better done with a few three-by-five cards or a pencil. The 'I've got it so I'd better use it' notion often leads to trivial or inefficient uses." (This is from John Bear's wonderful book, *Computer Wimp* [Berkeley: Ten Speed Press, 1984], which I recommend to all technophobes.)

Since this book is so directed to your efficiency, I feel obligated to render warning upon warning about computers before showing how they can save time for you—for some of you, and by now you know who you are.

Many people purchase a computer to speed their handling of family finances. To that idea, *Money* magazine says, "Family finance programs generally either are unnecessarily complex or simply duplicate what you can do with a thirty-dollar calculator and a little patience. . . . Probably in less time than it takes to set up the budget program and tally monthly totals for income and expenses, you could do the same job with a calculator and a ledger." Writing in *The Atlantic* in 1982, James Fallows admitted, "At the end of the year, I load the income-tax program into the computer, push the button marked 'Run' and watch as my tax return is prepared. Since it took me only about six months to learn . . . I figure this approach will save me time by 1993."

One more word of warning, especially if you're a

business owner in search of ninety-minute hours: computers are so much faster than humans that faulty computers have ruined many a business before anyone knew something was wrong.

IF NOT TO SAVE TIME, WHY DO PEOPLE USE COMPUTERS?

To warn you away from computers one final time before inviting you to use them, I want you to gain the insights that John Bear obtained interviewing hundreds of people while writing his book *Computer Wimp*. He discovered that, in order, the ten most popular uses for computers were

1. Games—not really timesaving
2. Electronic filing cabinet—also not timesaving
3. Personal finance—nice, but not timesaving
4. New skills—lessons that teach, but don't save time
5. Trivial uses—recipes, comic-book collections, and more
6. Word processing—a real timesaver, even more than a pencil with an eraser, my first real word processor
7. Business use—accounting and the like, a true timesaver
8. Research—tapping into data banks, a major timesaver
9. Practicing for the office—sort of a security-blanket type of use, not really a timesaver
10. Cuddling—owning the newest novelty, not a timesaver

Now that you're so well versed in ways a computer will *not* save time for you, let's examine some of the ways it will.

SAVING TIME BY GAINING INFORMATION

If you need to do research, gather data, or compile information, and you had the foresight to invest in a computer with a modem (modulator/demodulator), you can use your phone and computer to hook up with giant computers in other parts of the world, sending and receiving information over the phone lines. These giant computers or data banks—such as The Source, Scripps Howard News Service, Dow Jones News-Retrieval, and CompuServe—enable you to tap into their resources for old or new newspaper stories, book reviews, travel reservations, stock market prices, brokerage services, weather data, on-line shopping, movie reviews, bulletin board services, and other sources of information and data.

There are almost three thousand electronic data bases currently in existence, a data base being a collection of information related to a specific topic. Don't think you may have to subscribe to all three thousand. A subscription to one usually provides you with a subscription to many. It's like having a library at your fingertips and is incredibly timesaving if you need the data. To gain the benefit of these data banks, you first subscribe, then you pay an hourly fee plus phone charges. The prices are very reasonable if your needs are acute.   ‚

Heavy-duty researchers say it takes about forty-five minutes to make the rounds of all the state and national wires and to read *The Wall Street Journal.* You could get all that information, without ever touching a computer, simply by subscribing to *The Wall Street Journal.* But the difference here is timeliness; you can read tomorrow's *Journal* on your computer at 6:00 P.M. tonight.

You can also save scads of time by having your computer identify your areas of interest, so that you don't have to read reams of electronic data to find the gems you require. By typing in codes, you can select items to focus on, such as certain companies, industries, geographic regions, baseball teams, governmental departments or agencies, and the like. Your computer then pulls up headlines or stories dealing with those subjects.

Stock market investors need up-to-the-second data on the performance of certain stocks. Jim Cameron, a New York news consultant to radio stations, is also an avid stock market player—and a subscriber to both CompuServe and Dow Jones News-Retrieval. "I'm an investor," he said. "And what an investor needs is up-to-the-minute information about what his stocks are doing. . . . I figure I'm ahead of about 90 percent of investors because I can access stock market information instantaneously." No question that the speed of a computer aids and abets the goals of Mr. Cameron and others like him.

You can read the information you access on your screen, or have it printed on your printer. For instance, CompuServe offers the AP's Executive News Service, which is a clipping service that flags articles on selected subjects and stores them in electronic files until you have a chance to read them. Devoted ninety-minute-hour disciples might subscribe to NewsNet—a service that provides the full text of nearly three hundred newsletters.

Any chapter about computers and time ought to have a few words about computer printers—so here they are:

Laser printers provide you with more ninety-minute hours than standard mechanical printers because they print so fast. They also cost a lot, but how much is timesaving worth to you? Laser printers are also quieter than mechanical printers. So if you buy one, you get speed as your prime benefit, relative silence as an extra. Anyhow, there are nonlaser printers that print almost as fast as laser printers but not nearly as quietly.

I have a fast mechanical printer; my NEC printer does not know the meaning of malfunction. But I practice my ninety-minute-hour preaching by delegating my printing of lengthy documents and manuscripts to a fellow with a laser printer. This gives me the savings of time I need, and I don't mind the economy either.

SAVING TIME WITH WORD PROCESSING

There is no question that my computer saves me scads of time in word processing. Recently I wrote a direct-mail letter for a large bank. Because we were making the offer contained in the letter to nine different subgroups, I had to write nine versions of it. I saved scandalous amounts of time in two ways:

1. I wrote only one version of the letter, then drafted eight separate paragraphs aimed at the subgroups, inserting them instantly into the computer. The result: nine letters in the time usually necessary to write two.
2. When the bank's legal department got their hands on the letters, they made their usual changes, and the nine letters came back to my desk. Generally, incorporating those revisions would entail at least

one day of rewriting. Instead, because I could insert only the changes into my word processor, the whole task took one hour. Yet I was paid as though it took a full day. On that one day, my whole computer system paid for itself.

For doing versions of a standard piece or for making revisions, a computer is heaven-sent. But to be honest, unless you spend one thousand or more hours a year at a typewriter, word processing may not be cost effective for you. It will save time for you. It will improve the quality of your life. And it will—really—be fun. But if you can delegate the typing to a typist or if you only spend a few hours per year at a typewriter, you might consider time-savers other than word-processing computers.

I am writing this book with my word processor. That will enable me to edit simply and quickly. The writing of a book is a good reason for doing word processing. But when I write a letter, I use my typewriter. Why? Because I won't have to do any editing. Rewriting will be totally unnecessary.

Some computers, by the way, famed for their abilities in some areas, are absolute losers when it comes to word processing. I won't mention names because the technology is changing so rapidly, but be sure to check on word-processing capabilities if that's what you'll be using your computer for.

All word-processing programs process words, that is, move them about the way cutting and pasting might do. But some do it considerably faster than others. It's worth checking. Some programs provide you with a dictionary, a spelling corrector, and a thesaurus. All three are true timesavers. If you've been using a typewriter all these years, try to get a computer keyboard that matches the keyboard you've been using. This will make your transition into computerville that much easier.

SAVING TIME WITH NUMBER CRUNCHING

I suppose we all entered the Computer Age inadvertently when we bought our first hand-held calculators. I still have mine. My computer probably performs all the functions of my calculator and fifty-five other functions that I don't even know exist, but I don't spend much time number crunching.

However, there are many people who work all day with numbers and have found that they can get a month's worth of work accomplished in a day with a computer. Electronic spread sheets let you do all sorts of fancy projections and "what-if" charts, and allow you to explore the unknown, numberwise.

If you spend much time working with arithmetic, you can race your way to the ninety-minute hour with a computer. As with other users of computers, you'll find speed to be the main advantage here. But don't downgrade the other advantage: increased financial gain. You can earn a higher income because you use a computer for the proper reasons.

SAVING TIME BY KEEPING RECORDS AND SUMMONING THEM FORTH

Almost anyone can maintain a list of one hundred names. But computers are just the ticket when it comes to maintaining a list of ten thousand names, adding to the list, subtracting from it, then summoning it forth at your whim.

And names are just some of the records that can be kept by computer. Understand that keeping records is just part of the timesaving ability of a computer. But it is a waste of time unless locating and summoning forth those records is just as timesaving. There are too many horror stories of businesses that fed all their data into their computers for record keeping, only to be unable ever to retrieve the data again. The owners of those businesses would not consider computers timesavers.

SAVING TIME PUBLISHING

If you haven't yet heard of desktop publishing, you may be the only English-reading individual on earth who is reading of it here first. Desktop publishing is a true time-saver if you can put the words into your word processor, add the graphics with your appropriate software, lay out the pages with your other software, then print up the requisite number of copies with your laser printer.

If you had to attend to these tasks before the days of desktop publishing, this computer application can be a definite timesaver for you. But otherwise, it is not. If you weren't going to go through all that rigmarole in the first place, and you're planning on doing it simply because you can, desktop publishing may be a time bandit for you.

I publish a newsletter. I do the word processing and love every word, every process. But it would be inefficient for me to go through the graphics and page layout work— even though my computer would allow me to do so. So I delegate that work to professional desktop publishers. Believe me, I'd love to do the graphics, and perhaps some-day I will. But right now I am much too involved with other things to become a desktop publisher.

SAVING TIME COMMUNICATING

First, in order to communicate, you'll need a modem, a computer, and a telephone. With this equipment you're in a position to take advantage of huge advancements in communicating electronically.

Whereas less than 5 percent of computer hardware expenses have gone into communications until now, in the next five years around 30 percent will be spent in that area. We're speeding headlong into the era of electronic communication. Already a bit more than one-quarter of all computers sold are equipped with modems.

Right now, there are a host of E-mail (electronic mail) services. Nearly 1.0 million computer owners subscribe to these services, which allow them to communicate using their computers. Many of those owners are companies, so the number of actual E-mail users is 5.5 million.

If you're a computer owner who is into communications, you've got a computerized mailbox. The E-mail services can access your mailbox instantly if their communications software is compatible with yours.

It makes no sense to subscribe to an E-mail service if you don't do much communicating. But if you do a lot, you can save a lot of time. Many people consider E-mail an efficient alternative to the overburdened U.S. Postal Service. Marty Winston, a PR man from Texas, sends more than two hundred messages a week. You can be sure that E-mail is a major timesaver for him.

As I write this, the nation's 5.5 million E-mail users send 756 million messages a year. The U.S. Postal Service delivers a mere 140 billion.

Only small numbers of people can benefit from electronic communication right now because of the lack of compatibility of communication equipment and software.

But by the mid-1990s that incompatibility will have disappeared, thanks to a compatibility standard known as X.400—the key to a global postal system. Until then, you can only save time by communicating with computer owners who own systems that can communicate with yours. Or by investing in a fax machine. (More about fax machines in the next chapter.)

These days, a lot of computer owners who communicate electronically are forming SIGs, or special-interest groups. Hundreds of these groups now exist, organized around topics such as cooking, golf, business areas, marketing, and, of course, computers. These SIGs have their own message boards and libraries. On CompuServe, they hold on-line meetings in which participants type their comments at their computer keyboards. These conferences are almost always more efficient than face-to-face meetings, as you might expect—especially in terms of time. There is also considerable money saved by not traveling.

Ron Solberg, chapter president of the Public Relations Society of America in Chicago, says of the two hundred on-line meetings he's attended during the past three years, "There were about fifteen members of the technologies task force, and we did all our planning, assignments, and meetings on-line. It was just so efficient to be able to carry out business in this way, especially since it involved people from all over the United States."

Sounds ninety-minute-hourish to me.

If you are already saving time with a computer, you ought to give serious consideration to upgrading to a hard disk. Although this not-very-pretty piece of equipment runs anywhere from a thousand dollars up, it saves a lot of time because you don't have to handle a bunch of floppy disks.

One hard disk is all you'll need. And you'll have access to great amounts of information on that one disk, which incidentally doesn't look as much like a disk as a box. I use one hard disk and one floppy disk.

Perhaps you're a frequent flier and a person who honestly needs to use a computer to save time. Then you ought to be shopping for a lap-sized computer. You can use that instrument, along with your handy modem, to communicate almost anywhere—it's like carrying a portable office.

Thinking even smaller, you might invest a bit under three hundred dollars for a hand-held computer that serves solely as a computerized dictionary-thesaurus. It defines 80,000 words, based on *Webster's Third New International Dictionary,* Collegiate Edition, copyright © 1986, and at the touch of a button gives you access to 450,000 synonyms. The manufacturer says it lets you look up words five times faster than a standard dictionary or thesaurus. If you are constantly consulting these reference tools, this is a computerized godsend. If you rarely peek into those volumes, it's a senseless purchase.

COMPUTERS AND YOU—A MARRIAGE MADE IN HEAVEN?

Computers often save far more time in theory than they do in real life. There are more than a few tales of large corporations that became fascinated by the speed, capacity, memory, and computational abilities of some new computer. They invested heavily in a new-generation computer, and a frantic search ensued for things for the computer to do. Rather than admit that the powerful computer was not much more than a high-tech white elephant, the corporate types invented work so that the computer ended up turning out endless reams of information that nobody really wanted, not many people needed, and few if any people could use.

Ownership of the latest fashions in computers had become the end. Saving time certainly wasn't factored in. Sadly, nobody ended up with the right information; nobody ended up saving any time whatsoever.

Money managers would argue that the capital investment is so high that computer ownership can be justified only if the computer is being used all the time. The truth is that generating worthless information is far more expensive and inefficient than not using the computer at all.

I hope that you have been alerted to the potential errors in purchasing computers, as well as the timesaving benefits that come to those who have justifiably made a purchase. There are consumers such as the banker who said, "I have a portable lap-top computer, and it goes where I go. My computer is an extension of me, and when I'm on-line, I extend even further. My office is no longer a place: It's a state of mind."

Happily, I can say that my computer, while hardly an extension of me—and thank heavens for that—has expanded my capabilities and income while saving me a bundle of time. I hope yours can do the same for you. If you're not sure, *caveat emptor computerum.*

9

Extending Time with a Telephone and a Fax Machine

During the course of my work, I often make telephone calls that require me to have reference materials handy. While on the phone, I may have to consult information from file data, books, photos, my wife, my Rolodex, my daily calendar, my computer, and several other items in my office or house. But this chapter is not about those types of phone calls. This chapter is about phone calls that require no specific reference materials—the calls where you can wing it, relying strictly on your mind.

You can make those phone calls, saving impressive hunks of time, by the use of a car telephone, an air phone, or a mobile phone. Depending on how much you use the telephone, these types of phones and nonreference-material types of phone calls make a terrific timesaving combination. They are the very essence of time extension and the ninety-minute hour because they allow you to do one thing, make your phone call, while you are doing something else: driving, flying, or even walking in the park.

IF YOU CAN PUT IT ON PAPER, YOU CAN FAX IT

There is another timesaving method of using the telephone line, but not the telephone, that you probably know about and may even own or lease already. It's the phenomenon called fax—which is short for facsimile. This is not a computer technology, which is why you didn't read about it in the previous chapter. But it is one of the fastest-growing segments of the timesaving and office automation industries. It's also appearing in a multitude of homes, especially those with home offices.

The facsimile machine combines the accuracy and the tangibility of the written communication with the speed of the verbal. A Minneapolis corporate executive said, "I used to waste a lot of time trying to reach people on the phone. Once I finally got them, I'd use up maybe an hour a day shooting the breeze with them about how the weather was and how their local sports teams made out the night before. I don't waste as much time on the phone now since I got a fax."

A New York jewelry dealer adds, "One of my people can design a piece of jewelry in my office in New York and we can fax the design to my factory in Hong Kong in about thirty seconds. There's no mail to worry about, no overnight express."

And finally, a Los Angeles magazine publisher tells us, "We were on deadline and needed a photograph of a prominent Japanese industrialist in a hurry. There wasn't one available in Los Angeles. Finally, we contacted his company's headquarters in Tokyo. A perfectly usable photograph was sitting on my desk inside of twenty minutes."

Using standard telephone lines, the facsimile machine instantaneously sends and receives any kind of information that can be assembled on paper. While this goes on at lightning speed, the original document remains intact, and the reproduction is perfect.

If you think you're hearing more and more about fax in America, you should hear about the popularity it has achieved in Japan. There are about four times as many fax installations in Japan as in the United States.

Office fax machines can be leased for less than a hundred dollars a month. Purchase prices have dropped to less than one thousand dollars, and will eventually run under five hundred dollars. For the advanced model, using plain, not thermal paper, figure on a price tag of five thousand dollars. Fax transmissions cost about the same as ordinary phone calls. Just ask the Philadelphia businessman who claimed, "We used to spend almost a thousand dollars a month on messenger and express delivery services. Now I can fax the same fifty documents a month for less than fifty dollars."

Fans say that by the mid-1990s fax machines will be standard equipment in homes so parents can fax the kids' art projects to adoring grandparents. Another plus for the fax is that it is a cinch to operate, making it perfect for computer-shy executives. Consider the Dallas executive who said, "My five-year-old daughter is capable of running my office fax machine for me. One of the nicest things about fax is that no company needs to spend time and money on training courses."

Another advantage: The new fax machines will call back busy signals at regular intervals; if the line is still busy, the fax reminds you to keep trying. Fax machines can be preprogrammed to send to dozens of destinations regularly at times when phone rates are lowest.

Unlike computers, for which the communications software must be matched, fax units need not be the same brand and model to send to and receive from each other. A small business or self-employed entrepreneur can invest in a single fax unit to communicate with customers, prospective clients, and suppliers. New fax machines can even eliminate the need for paper. They can transmit an instant TV freeze-frame, which is received on a machine also equipped with a TV screen.

Fax machines are now considered the quickest and most inexpensive way to send written communications. They take less than a minute and cost less than forty cents (even less at night). And it takes only fifteen seconds per page to send an exact copy anywhere on earth.

There is even a fax telephone directory. The Official Facsimile Users' Directory runs sixty-five dollars and is available from FDP Associates in Manhattan (212-503-4100). The current directory lists 120,000 fax numbers in the United States, Mexico, Canada, Switzerland, West Germany, France, and the United Kingdom. This represents only 10 percent of the 1,200,000 fax machines in the United States. On the other hand, it represents 90 percent of the machines used by the Fortune 1000 companies and

86 percent of those at major U.S. law firms.

Fax machines are now selling at the rate of 275,000 per year and are expected to grow fivefold by 1993. Although sales are now in the domain of big businesses, about half of the small businesses in the United States will use fax machines by 1993. Already, the industry's annual revenues are over $1 billion.

(This may spawn a generation of fax junk mail, although legislation against it is being considered. Fax owners have to pay for the paper used, and they don't want to pay for unsolicited sales messages.)

The smallest fax machine actually fits in a briefcase. As in other electronic breakthroughs, look for the prices and sizes of fax machines to get smaller and lower as time goes on. I own one and I wish everyone I know owned one too. Eventually, most everyone will.

OFFICER, FOLLOW THAT CAR PHONE!

Many people with car phones claim to get a lot of real business done while driving. Instead of tuning in to their favorite deejay, these ninety-minute-hour devotees are dividing their phone calls into two categories:

1. Those that must be handled while surrounded with the reference materials I mentioned and thus must be made from office or home;
2. Those that need no reference materials and can be handled from a moving car.

Cellular telephone technology has enabled 1.8 million motorists to complete important phone calls while driving all over an urban environment, where the proper cellular relay equipment checkerboards the metropolis. These days, that equipment is in almost every large city and most medium-sized cities. And a cellular telephone enables you to call anywhere in the world where there's the right equipment—a lot of places. So if you're thinking in terms of calling range, you're a few decades behind the technology.

Cellular phones may be purchased, leased, or leased as part of a purchase plan. They may be installed in your home or office. Huge international corporations as well as start-up corporations are getting into the act.

Modern cellular phones make it possible for motorists and their passengers to make the calls simply by pushing a button since the new phones have memories that store frequently called numbers. You no longer even have to hold a receiver in one hand and a steering wheel in the other.

The disadvantages to this technology—other than the fact that you might become so engrossed in the call that you drive off a cliff—are that you are unable to take notes (safely) and unable to refer to existing material for needed data. But it's pretty simple to get around those obstacles. And when you do so, you'll find yourself in control of more ninety-minute hours than you might have imagined.

Sure, there's a tiny loss of fidelity in the sound transmission, but there's a glorious gain in time saved, in convenience, and in the intelligent utilization of time.

If you have a sixty-minute drive and you can handle thirty minutes of phone calls during that drive, you have lived a pure ninety-minute hour. If you can make sixty minutes' worth of phone calls—each one justified by the cost per minute, you're up there in the rarefied air of the one-hundred-twenty-minute hour. Naturally, you save no time if you use your phone just to make contact or to say hello or to tell your friends that you are calling from your car.

Beware of quality when purchasing your car phone. The prices are dropping, but car phone owners claim you get what you pay for. The price range is from about eight hundred dollars to about three thousand dollars, but I've been warned away from the very cheap variety by more than one person. As this book goes to press, monthly rates range from twenty-five to forty-five dollars—this is in addition to the purchase price—and you pay forty-five cents per minute for calls from 7:00 A.M. to 7:00 P.M., twenty cents per minute for calls after 7:00 P.M. and before 7:00 A.M.

I can think of few situations as horrendous as getting in my car to escape into the natural beauty of the countryside when suddenly my car phone rings. "Lynch the inventor!" I think at the time. But you don't have to give your car phone number to anyone if you don't want to. And this capability of being summoned while in the car is a major plus for some gung-ho timesavers. They tell their secretaries, "Call me the moment Mr. Bigbux calls. I'll be in my car driving out to see Ms. Maxibux."

As with a computer, it doesn't make sense to own a cellular telephone for minor conveniences. But if you must spend time on the phone and must travel by car, the new cellular telephone technology is perfect for timesaving.

It's not difficult to weigh the cost of owning a car phone against the profit opportunities it presents. You can get a clear idea of whether to invest in a cellular telephone by answering these five questions:

1. How much time per day do I spend on the phone?
2. How much time per day do I spend in a car?
3. Can I make the type of calls suited for car phones, the kind that do not require the physical presence of reference information?
4. Is the cost saved by my making the calls from the car greater than the cost would be if I didn't have the phone and made the calls from somewhere else?
5. Is driving and phoning simultaneously the best possible use of my time?

You may decide to do business while driving around making or taking calls. You can always keep reference materials in your glove compartment, attaché case, or trunk.

One final consideration: If you have great difficulty driving and maintaining a conversation at the same time, *don't* buy a car phone. It's not worth the time you may spend recuperating from an accident!

SAVING TIME IN THE FRIENDLY SKIES

If you might benefit from the use of a car phone, you might also benefit from an Airfone. These conveniences are becoming more and more commonplace, especially on larger planes. You simply insert your credit card into the phone unit, take the phone to your seat, and make all the calls you want. The sound transmission on airplanes isn't great, but I hope it will have improved by the time you read this.

That static-ridden sound didn't stop me from making

five calls last month as I flew from coast to coast. They cost a $2.00 setup charge per call. Plus $2.00 per minute. I'm sure I earned more than five times that cost with one of those calls. Can't say that about the ones to my mother, mother-in-law, wife, and sister. But then, how can you put a price tag on being able to say, "Happy Mother's Day" from 39,000 feet up?

The whole communications world is acting on the newly perceived value of time. That's why there are telephones all over the place. Amtrak offers Railfones, though only on the New York to Washington Metroliner as of now. They work in much the same way as Airfones do, except that they don't work while in a tunnel, something you hope you never learn with an Airfone.

Car rental firms also have car phones. Budget already is leading the pack with car phone installations, charging three dollars per day for the phone. Local calls are billed at ninety-five cents per minute. And car renters can make calls to or receive calls from anywhere in the world.

Planning a visit to Houston soon? If so, you'll be in the first city to have installed cellular pay phones in their taxicabs. These phones even have their own teeny-tiny meters. Cost: a buck a minute, payable to the cabbie. The dialing range is six hundred square miles, which just about covers Houston's local dialing area. You can expect taxi phones in most large cities by the mid-1990s.

Mobile telephones which fit into attaché cases are also in increasing use. Portable cellular phones employ battery packs, enabling you to use them in your car, in a rental car, on a boat, on the job site, anywhere within the cellular reach. I once took a plane trip from San Francisco to Atlanta, no short hop, and watched a man on the phone the entire flight. From the overheard snatches of conversation and the note taking I witnessed, I could see that the man was conducting business. Instead of watching the movie, reading a book, or sleeping, he turned a usually wasted travel day into gold.

The secret of using this sky-high-tech form of communication is to plan your calls. Have the necessary reference materials at the ready, and be sure you've got your book of phone numbers along. It also helps if you can alert the person you'll be calling in advance. Getting a busy signal or a wrong number can happen, and they cost the standard Airfone rate, which happens to be double the rate for operator-assisted international calls.

MORE SHORTCUTS TO NINETY-MINUTE HOURS

The Sharper Image (800-344-4444) offers few timesaving devices that work on jet planes but serves up a gaudy array of timesaving phones and answering devices in all their catalogs. A recent edition offered a hands-free telephone that looks like a headset. The description of the item said that it enhances efficiency and improves business productivity. The Sharper Image and The Price of His Toys (800-447-8697) are worthwhile catalogs for any ninety-minute-hour fan, because they give a good overview of the state-of-the-art communications goodies for timesavers. And *The Catalog of Time-Savers,* available free by calling 800-748-6444, offers a myriad of timesaving devices plus subliminal tapes and a whole lot more.

Coming up on the communications horizon to save time for you (but not as much time as car phones, Airfones, Railfones, or mobile phones) are E-cards, developed experimentally by AT&T. The cards are for use only at specially equipped public phones. Each card will have built-in microchips that speed-dial up to fifty phone numbers, store medical records, incorporate an electronic datebook and reminder pad, and even split call billing among as many as eight accounts.

Speaking of things tiny, a company called Metagram (800-262-6382) offers a pocket-sized message terminal on which a full typed page's worth of information can appear. When you hear a tiny beep or see a flashing light, just read the message that has been electronically relayed by a caller. Metagram message systems work seven days a week, twenty-four hours a day. By saving you the time it takes to receive info by phone, this mighty message mite earns its way into the ninety-minute hour. The unit costs less than five hundred dollars to purchase, less than twenty-five dollars monthly to lease (plus service).

U.S. Page, Inc., employs satellite technology in their paging system so that your customers, office, spouse, co-workers, or the IRS can reach you nationwide, toll-free, seven days a week, twenty-four hours a day, within thirty seconds of the time a phone call comes in for you. An LCD readout identifies the caller's twelve-digit message or phone number. U.S. Page tells us that the cost of their technology is a shade more than a dollar per day. Mobile-Comm offers a similar product called SkyPager. Their advertising claim sums up the benefit of such devices: "Out of Town. Not out of Touch."

Four grams is the total weight of LiteSet—which rests in your ear and extends toward your mouth, acting and sounding just like a truly convenient cordless telephone. It attaches to a push-button pad when you want to make calls. Once you've pushed the buttons, you're free to roam and your hands are free to do whatever you wish. Happily, your shoulder is not scrunched up to keep communications open. Plantronics, Inc. (408-426-5858) sells the unit for a bit under four hundred dollars. The noncordless variety is under eighty dollars.

A relatively old (mid-1980s) communications technology that seems to be zooming in popularity is voice mail. Since only one-fourth of business calls are completed on the first try, voice mail eliminates telephone tag. It's a high-tech networked answering service that enables you to receive messages, reply to them, send memos to individuals or an entire organization, update recorded greetings, and more—all in a single call. The system digitizes a voice message for storage on a computer disk. Callers are greeted with a message in the recipient's voice and are invited to leave a message. The recipient is then instantly alerted.

It seems as though the entire communications industry is going all out to give you your fair share of ninety-minute hours.

10

Creative Waiting to Gain More Time

Time is either spent or it is wasted. If you decide to do absolutely nothing with your free time, it is not wasted because you spent it just the way you decided to. But if you allow time to pass without spending or wasting it because you gave it no thought, that time is wasted.

If you think it's stupid to waste money, think of how much more stupid it is to waste time. Time comes in too limited a quantity to waste. Time is far too valuable to let pass without thought and purpose. Nonetheless, many of us find ourselves in situations that appear to be no more than a waste of time. Sadly, some of those situations *are* wastes of time, and there's little you or I can do about it.

WAITING IS ON THE UPSWING

You've probably noticed that waiting is becoming more and more popular. In addition to the old standbys—such as waiting for postal clerks, late airplanes, and doctors' appointments—traffic jams are more frequent, more abundant, and longer than ever.

If that weren't enough, telephone technology has inflicted upon humankind that condition that is best described as "permahold." I'm sure you know it well. You call a company, they greet you, then ask you to "please hold." Sometimes they play music to soothe you. Sometimes they connect you with a radio station so you can listen to music and commercials. These days, they might play a message loaded with reasons to patronize their business. But usually, they leave you with silence: the expectation and need to communicate on your end, utter silence on theirs. The silence lasts and lasts. Or the music lasts and lasts. Permahold. Wait it out or hang it up.

If you're a devotee of ninety-minute-hour thinking, while on permahold you might file or clean out your files. Or you might sort out or initiate correspondence. Perhaps you'll read a memo or write one. Maybe dictate one. Probably you've got a small stack of magazines and newsletters nearby that are reserved strictly for permahold situations. Now's an ideal time to browse through or selectively scan them.

All of these options are better than merely waiting it out. But they might not be better than hanging it up and getting on with some other chore. If you decide not to be victimized by waiting as so many people are, you'll use this time creatively rather than wastefully. You won't be a victim. Admittedly, you won't save giant chunks of time.

But ninety-minute hours are composed of both large and small segments.

Think of the places you might wait:

- Restaurant
- Airport
- Home
- Barbershop or hairstylist
- Train station
- Bus station
- Post office
- Bank
- Meeting
- Appointment with doctor, dentist, businessperson
- Traffic
- Supermarket
- Movie
- Retail store
- Gas station
- Garage

I have no doubt that you can add to the list. There doesn't seem to be a shortage of waiting situations. If you added up all the minutes you waited during the course of one year in the situations listed, you'd probably discover that you spend several days just waiting—especially if you include permahold.

But when you apply ninety-minute-hour thinking, many of those situations are salvageable. In the hands of a person who is intrigued with the idea of ninety-minute hours, many otherwise time-wasting circumstances can be transformed into valuable time spent.

There are ways to effect this transformation, methods for intelligently dealing with what you can reasonably foresee as potentially wasted time. Set your mind to take advantage of, even welcome, those moments when others twiddle their thumbs or stare ceaselessly at their wristwatches.

In Robert Heinlein's fine science-fiction novel *Stranger in a Strange Land,* the visitor to Earth from Mars has a healthy and laudable attitude toward waiting. He accepts it. He does not resent it, curse it, or avoid it. Instead, he understands and says that "waiting is." And whatever it is, it is undeniably a consumer of time. But that does not signify evil.

In fact, to the Martian visitor and to ninety-minute-hour buffs, waiting is a granter of time—an opportunity to accomplish or to relax. Ninety-minute-hour thinking is that which welcomes waiting time, realizing that waiting can be fertile for time extension. But the fertility exists only after you have attended to the two waiting prerequisites.

YOU MEAN THERE ARE PREREQUISITES FOR WAITING?

You bet your stopwatch there are. It is definitely possible and obviously advisable to turn humdrum waiting into creative waiting—the kind dictated by the value of time. Creative waiting is spent with wisdom and insight to accomplish, achieve, or rejuvenate. But creative waiting has two prerequisites:

1. PLANNING

You've got to plan ahead for how you will use the time, depending on the type of waiting time you'll have at your disposal. I'm not talking about a written, formal plan. I mean a mental strategy, a realization of the opportunities that time might offer you.

2. EQUIPMENT

You've got to outfit yourself with the proper tools for creative waiting. Your gear, all tiny and lightweight, might consist of a combination of old-fashioned stuff, such as a small pad of paper and a ballpoint, along with new-fashioned electronic devices, such as a headphone cassette player-recorder, calculator, maybe even a mobile telephone.

Not only have you got to prepare your mind to transform waiting time into useful time but you've also got to give yourself access to whatever work or equipment you'll need.

A BRIEF WORD ABOUT PLANNING FOR WAITING

Look over your tasks for the day or the week. Consider work tasks and nonwork tasks. Many obviously cannot be handled while waiting for an airplane or at a long red light. But a few might be ideal to categorize as reserved for creative waiting opportunities. Author Scott Turow wrote the bulk of his best-seller *Presumed Innocent* in longhand in notebooks while commuting.

To add creative waiting to your timesaving arsenal, you'll need a container for your tools. Probably, that place will be a briefcase or attaché case. In it will be a folder of paperwork chores that can easily be accomplished while waiting. Don't forget, this includes work tasks as well as personal tasks.

Your planning will also be the reason your case includes appropriate reference materials: address book, relevant past correspondence, blank paper for writing—notebook or notepad—stationery, stamps, envelopes, plus anything else you might kick yourself for not bringing.

Now you are prepared for creative waiting. Your carrying case is poised and ready to do whatever you need to do. Perhaps it is a special case, reserved exclusively for while-u-wait tasks. But more likely it's a section in your regular case with the plans, papers, and paraphernalia for enlisting the aid of waiting time for your cause.

Your mind-set is different too. You view chores not just as work but with the fine tuning that tells you whether they mean work that must be accomplished in a certain setting or work that can be attended to virtually anywhere.

The more places tasks can be handled, the more time you'll save. By doing certain tasks while you would otherwise be wasting time waiting, you will accomplish more, gain time, and feel a lot differently about waiting. And let's not forget that you always have the option to use your waiting time to daydream, meditate, rest, relax, or stare off into space.

POCKET PARAPHERNALIA FOR THE NINETY-MINUTE HOUR

Tucked into your case, along with the pad of paper and pen, you may want to include a cassette player-recorder and headphones.

You might also be wise to carry:

- A microcassette player-recorder for dictation or listening. The electronic memo is part of the Information Age and a logical accoutrement for the ninety-minute hour

- A calculator, if you do any type of calculating
- Appropriate cassettes for hearing memos, instructional material, or subliminal suggestions
- Blank cassettes for dictating memos, letters, ideas, flashes of genius, notes to yourself
- A small but well-chosen selection of newsletters and other reading materials
- A mini–copying machine
- A lap-sized computer
- A beeper to connect you with your answering service, in case your waiting puts you in proximity to a telephone
- A portable telephone

If you think these ideas are advanced, my friend Neil travels everywhere with his modem—so that he can plug into any convenient push-button telephone, connect his minicomputer, and communicate worldwide. Because Neil has that contraption, he uses it. He knows that he will, but not necessarily when. Still, daily, he uses his modem for sending and receiving electronic memos.

THERE'S AN ELECTRONIC MEMO IN YOUR FUTURE

Electronic memos are bona fide timesavers. There's a clear correlation between the comfort you feel with dictating and hearing them and the time you save. Try to become adept at dictating quietly but clearly even in crowded areas. Become facile at saying notes to yourself, and you'll know just where to find them.

In time your electronic memo will connect up with a computer that has voice-recognition capability. Then, the computer can put your verbal notes into writing, even distribute them to the appropriate personnel. In fact, this technology already exists. And so, alas, do memos—and in approximately the same abundance as molecules of nitrogen, the most common element in our earth's atmosphere.

The electronic memo, combined with a voice-recognition computer, will be a welcome contribution to the ninety-minute hour because of its ability to condense time without sacrificing the good sense of putting things into writing. You can plug into that future before it zooms on by you simply by getting used to electronic memoranda.

LET YOUR UNCONSCIOUS HELP YOU WAIT

Your unconscious mind will soon realize that you have bestowed upon it a valuable tool for recording ideas and suggestions. Rather than merely thinking them and letting them float away on mental breezes, you'll become conditioned to saving them, storing them, building on them, accomplishing more as a result of them.

In time, your ideas will become limitless. After all, now you've got a place to record them. So what if you can't act on them now? You probably will later. If even one out of one hundred ideas you record on your pad or blank cassette is a biggie, your life can take a dramatic turn for the better.

Once you start to comprehend the possibilities that come from keeping your entire mind available for tapping at any time, you'll want a device for recording your thoughts. Whether you choose a pad of paper and pen or a microcassette player-recorder, buy several. Keep each in a different place, such as

- Your pocket or purse—available whenever you're waiting

- Your car—just in case your device wasn't in your pocket or purse

- Your nightstand—because good ideas don't care where or when you get them

- Your briefcase or attaché case—the ideal place for all your ninety-minute-hour equipment

FIRST THE MIND-SET, THEN THE MATERIALS

The benefits of waiting come only to those seekers of the ninety-minute hour who prepare themselves. Although scheduling your tasks is the most effective way of dealing with them, allow flexibility in that scheduling so that you can take advantage of waiting time to handle chores scheduled for another time.

Some human behavior specialists recommend that you save your stickiest problems for periods of waiting, when you are limited in time and pressed to come up with a solution. Those same experts tell us that waiting time is "newly created" time, time that comes unexpectedly. Such time is especially fruitful for solving tough problems. That's probably because newly created time is ripe for newly created solutions.

EVEN A MARATHON STARTS WITH BUT A SINGLE STEP

It may be impossible to utilize all your waiting time effectively, but you can use some waiting time to make inroads toward ninety-minute hours. As I've said before, all work is constituted of starting and completing. By using your waiting time creatively, you'll be able to do a healthy amount of both.

Now you know how to wait without wasting time. You may never have thought you'd see the day when you'd welcome the traffic coming to a halt as a slow freight train lumbers across the road, but perhaps that day is coming.

11

Accessing Your Unconscious to Gain More Time

The tools of the ninety-minute hour come in several forms. The tool of delegation is an *art*—the art of selecting which tasks to delegate and to whom to delegate them. The tools of audiotapes and cassette player-recorders are *technologies,* employed for years in directions other than the primary goal of saving time. The tool of subliminal suggestion is a *science.* Television is another technology, as are computers and telephones. The tool of creative waiting is a perception, an *attitude.*

You have at least one more tool in your bag of time-saving tricks. I waited until now to write about it because I want you to realize its power compared with that of the other tools; it can contribute mightily to your success at gaining extra time for yourself. And that is the aim of this book.

The tool I call to your attention is your own mind—more specifically, your unconscious, which constitutes nearly 90 percent of your brainpower. Philosophical psychologists might argue that you are a tool of your unconscious rather than the other way around. However, for ninety-minute-hour purposes, the unconscious is your tool.

Of course, you've been using it all your life. But you've rarely activated it by conscious action. As a time-saving tool, your unconscious can prove an invaluable aid—if you know how to communicate with it, if you know how to access it.

You may have ways to stimulate your unconscious to help you attain your goals. Possibly, they include one of these methods:

1. Affirmations
2. Visualizations
3. Self-hypnosis

If you are convinced that you can get more out of life by gaining extra time, and if you are willing to use the tools of time extension, then the cooperation of your unconscious can unite all your efforts to give you the benefits of the ninety-minute hour.

YOU CAN PROGRAM YOURSELF FOR THE NINETY-MINUTE HOUR

To enlist the considerable help of your unconscious, you've got to take regular conscious actions—less than five minutes' worth a day—until your conscious desire to gain extra time becomes an unconscious behavior pattern. When you've put yourself on automatic pilot for the ninety-minute hour, you need no longer consciously access your unconscious.

Excuse me for engaging in all this psychobabble, but it's important that you understand how to program yourself to save time. Let me exchange the language of psychobabble for the language of technobabble to show how you can put your unconscious mind to work for you.

Your unconscious will serve as the computer and memory. Your conscious will serve as the keyboard for entering ideas into your computer. Your affirmations, visualizations, and self-hypnosis will serve as the software programs. Ninety-minute hours will be the printouts, the solutions you sought from your computer.

Not only will you delegate well and make intelligent use of audiotapes, subliminal suggestions, and the other ninety-minute-hour tools, but because your unconscious mind is so powerful, so comprehensive, it will see, invent, and incorporate the use of still more tools—timesaving modes of behavior that will give you even more extra time.

AFFIRMATIONS START OUT TO BE LIES, END UP AS TRUTHS

One of the most potent, proven, and increasingly utilized ways to access the unconscious is affirmations. An affirmation is a statement that you wish to be true, but that really isn't true yet. There are six things you should know about affirmations if you want yours to transform from affirmation to reality:

1. Affirmations are positive statements.
2. Affirmations are so personal that they contain your name.
3. Affirmations state hopes as though they are facts.
4. Affirmations come true faster if you say them aloud to yourself twice a day—just before going to sleep and just after awakening. That's when the unconscious is most receptive to suggestions made by the conscious part of your mind.
5. Affirmations are far more likely to become reality if you put them into writing. Select just one affirmation, then write it out in longhand twenty-one times three or four times each week. If you have more than one affirmation, that's fine. But put no more than one into writing on the days you write it out. That will take about three minutes if you write legibly. And you should.
6. Affirmations don't come true unless you keep them to yourself. After they've come true, you can tell them to anyone you want, but until they do turn into reality, mum's the word. So if you share your bed with someone, you've either got to whisper your morning and evening affirmations or you've got to say them alone in a different room.

A typical affirmation attuned to the ninety-minute hour might be: "I, Glenn, have two extra hours per day each day."

The first few times you say your affirmation, you'll feel like an idiot. But I give you my solemn word that they do work. When you are practicing an affirmation, you are making a positive statement of a hope that you wish to be a fact, and you are stating it as a fact even though you know it is only a hope.

Your unconscious mind, however, does not know the difference between a fact and a hope. It believes everything it hears. And twice a day it hears that you've got two extra hours per day each day. Generally, somewhere around thirty days from the time you start, you will realize that your affirmation is coming true. What has happened is that the titanic power of your unconscious has actually rearranged your behavior to transform your hope into a fact.

It dawns on you that you are using as many tools of the ninety-minute hour as possible. You are becoming the soul of efficiency. Better yet, you don't feel stressed out. In fact, you feel less stress because you have more time—two extra hours a day every day. You'll be astonished at the simplicity as well as the brute force, of affirmations. It wouldn't surprise me if you soon develop another set of affirmations designed to solve other problems, achieve other goals.

Just don't forget that affirmations are to be shared only between your conscious mind and your unconscious mind. If anyone else gets in on the affirmation, the power goes out of it. Sorry, but that's the way it is.

AFFIRMATION HINTS

Hint A: A couple of times a week, say your affirmation while facing a mirror, looking directly into your eyes.

Hint B: A couple of times a week, say your affirmation three ways:

1. "I, Glenn, have two extra hours per day every day." This is you telling yourself about your extra time.
2. "You, Glenn, have two extra hours per day every

day." This is someone telling you to your face about your extra time.

3. "He, Glenn, has two extra hours per day every day." This is someone else talking about your extra time.

Hint C: Three or four times a week, write your affirmation twenty-one times. Write it seven times as though you are saying it to yourself, seven times as though someone is saying it to you, and seven times as though someone is saying it about you.

If you do your affirmation this way, you'll begin to "see" the hope as a fact. To "see" it in even clearer detail, try a little visualization.

VISUALIZATIONS ARE METHODS FOR SEEING INTO YOUR FUTURE

As long as you're going to take a few minutes a day to affirm your extra time, it makes a lot of sense to take a few extra seconds to ensure that that time will be yours.

Spend those extra seconds visualizing yourself with the extra time. Visualize yourself spending the free time, for work or leisure. Visualize the methods you'll use to gain the time. See yourself driving while hearing an audiotape. See yourself working with soft music in the background, implanting subliminal suggestions. See yourself asking an associate to handle a task for you. See yourself zapping commercials on your VCR, talking on your car phone. While you are seeing your affirmation as truth, *feel* it as truth as well. To use your unconscious mind for maximum effectiveness, enlist all three of your sensory systems: auditory, visual, and kinesthetic.

The best time to do your visualizing is when you are doing your affirming. When you say the words, the pictures will come to mind. If they don't, make a serious attempt to bring them to mind. It does not take long to visualize if you are affirming. One or two minutes each morning, one or two minutes each evening—to do both the affirming and the visualizing.

If these sound like New Age psychological techniques, that's because we are right in the middle of the New Age, and psychologists are just beginning to flex their cerebral cortexes.

VISUALIZATION HINTS

Hint A: Don't visualize in any state but a relaxed and untroubled one.

Hint B: Visualize as many details as possible: what you are wearing when you delegate work, including colors and all other details; the exact audiotape you'll be hearing in the car; what you'll do with the spare time you come up with on a weekend or an evening. The more you see in your mind, the quicker you'll be able to see it in real life.

Hint C: Because your unconscious mind is stimulated more by repetition than by any other single means, continue your visualizing regularly until what you are visualizing turns into reality.

These two processes—of affirming and visualizing— will take you a total of no more than two minutes per day. Toss in the mirror and the writing, and you've added about ten more minutes per week.

SELF-HYPNOSIS SOUNDS COMPLICATED,
BUT IT ISN'T

If you do it right, you'll rarely spend more than two minutes at a time self-hypnotizing—and you'll do it about three times a week.

The whole science of hypnosis used to be shrouded in mystery and hocus-pocus—typical of a science that is not yet understood by humankind. But that view is rapidly coming to an end. The U.S. Supreme Court recently ruled for the first time to legitimize hypnosis in legal proceedings. In fact, these days executives, homemakers, even students are practicing the simple art of self-hypnosis to achieve their goals. They realize that hypnosis does not control them; it enables them to control their unconscious minds to a degree. Because it does, it is an extremely valuable tool.

Self-hypnosis is a process during which you make reasonable suggestions to your unconscious mind. After you stop being self-conscious about self-hypnosis, you'll have a shortcut that leads directly into the most powerful part of your brain. And although I hope you begin to use self-hypnosis as a method to gain the maximum number of ninety-minute hours possible, I'll bet that eventually you'll begin to use the procedure in other aspects of your life. Many professional athletes use it to enhance their performance. Many millionaires use it to increase their fortunes.

Hypnosis, rather than the black magic it was once perceived as, is really another name for suggestion. In fact, the dictionary defines hypnosis as "an artificially induced state resembling sleep, characterized by heightened susceptibility to suggestion." Self-hypnosis is that same state—only with you making the suggestions.

Dr. Freda Morris, author of *Self-hypnosis in Two Days*, says, "If you want to take charge of your whole mind, your whole life, hypnosis is the tool of choice." She tells us that

> hypnosis is a state of consciousness in which you are more than ordinarily receptive to suggestions, can think clearly, can see vividly with your mind's eye, can feel intently, and can really listen to your deeper, wiser self. In this state, you have a unique control over your mental processes, your emotions, and your attitudes. It enables you to focus where you want to and to be calm, peaceful, concentrated, clearheaded, alert, totally engrossed in the mental activity of your choice, and completely undistracted by outside interferences and irrelevant thoughts.

In spite of the benefits of hypnosis, Dr. Morris cautions us that hypnosis is of no particular value by itself. She stresses that it is the way hypnosis is applied that makes it useful. I am suggesting that you apply it to implant in your mind not only an awareness of time and efficiency but also the proper behavior to save that time and gain that efficiency. This will happen because during your self-hypnosis sessions you will make suggestions that will actually alter your behavior after the sessions. You've probably heard of the phenomenon called posthypnotic suggestion. It's a power you can grant yourself through simple self-hypnosis.

Here's how to go about self-hypnosis: First, find a suitable spot—possibly even your office—where you can be assured of at least five minutes of freedom from interruptions. Then relax, make yourself comfortable, loosen any constricting clothing, uncross your legs, take three deep breaths, slowly close your eyes, and silently count from one to five, believing that with each number your unconscious mind will become more and more receptive to suggestions. After you've reached the number five, think and feel your affirmation. Don't say it aloud. Just think and feel it a few times; three would be enough, though five would be better. Speak silently to yourself about your goal of freeing up more time, concentrating on the ways to accomplish that goal.

Be sure you focus completely on the saving of time while you are thinking and feeling it. Make suggestions to yourself to streamline your behavior, to increase your effectiveness, to utilize your time with intelligence, to avoid wasting time, to combine activities whenever possible, to utilize planning and prioritizing as timesaving measures, and to establish any other behavior patterns that will help you in your quest for ninety-minute hours.

These are reasonable suggestions. Your unconscious mind will have no problem accepting them. Then, count backward from five to one, reminding yourself of how good you'll feel when you open your eyes. At the count of one, open your eyes, then go about your business. That's self-hypnosis. Doesn't take long. But don't let its ease and brevity fool you; it's a powerful ally.

SELF-HYPNOSIS HINTS

Hint A: Be sure you are comfortable and under no stress when doing your self-hypnosis. You want to be able to concentrate on the suggestion, not on your comfort or your problems. You want to focus all your mental energy on your goal of achieving more time for yourself.

Hint B: Don't worry about silence in your surroundings. Although noises may distract you at first, later you'll be able to tune them out. Your unconscious couldn't care less about what else is going on in the room around you.

Hint C: Remember that hypnosis is not a practice to be limited to professionals. It's just a fancy way of saying "suggestion." Once you realize that, you'll lose any lingering fears about hypnosis that you may have held.

Hint D: When practicing self-hypnosis, try to be rested but not lethargic, exercised but not to the point of fatigue, fed but not sated, and free from alcohol or drugs.

FOUR WEEKS TO PROGRAM YOUR MIND

Bottom line: Spend only four weeks investing thirty minutes per week (fourteen for affirming and visualizing one minute each morning and one minute each evening; ten for writing your affirmation three or four times a week and saying it into the mirror; plus six for three two-minute self-hypnosis sessions), and you can gain two extra hours per day.

Consider that you'll be able to save chunks of time like that for the rest of your life as you invest only thirty minutes per week for a few weeks. Sounds like a fair trade to me.

After these few weeks are over, your unconscious will move into gear, and you can get ready to enjoy your two extra hours per day. Then you can stop affirming, stop visualizing, and stop your self-hypnosis. Your unconscious will be properly programmed. You may gain a bit less than a full two hours each day; then again, you might gain more.

YOUR PLANS ARE YOUR GUIDE, NOT YOUR BOSS

Although I urge you to plan, I also hasten to point out that flexibility is a buddy to the saver of minutes. Don't be a slave to your list of assignments. Take them on when you feel you can do them best.

This kind of thinking seems foreign, especially to anyone who has never planned time, never tapped the unconscious mind. But the more you practice timesaving behavior, the more your unconscious will get the idea of what you are trying to accomplish. Even if you do zilch to enlist the aid of your unconscious, it will find ways to save time for you simply because it will recognize that you are finding ways on your own. But if you intentionally access your unconscious, using any one or all three of the techniques in this chapter, you will find that the ninety-minute hour is attainable.

ARE THERE ANY ENEMIES OF THE NINETY-MINUTE HOUR?

Several categories of people have been identified as the opposition to ninety-minute hours. Because it is so important to know thine enemy, I list a few here:

- Perfectionists
- People who love the sound of their own voices
- Long-winded writers
- People who are impressed with their own authority
- People who are in over their heads
- Egotists
- Bureaucrats

- Mind changers
- Overorganizers
- People without goals
- People without a lot to do
- Interrupters
- Decision avoiders
- The people in front of you at the automated teller machine
- People who put you on phone-hold for longer than thirty seconds
- People who call you (or have their secretaries do it), then put you on hold
- People who dissect the restaurant check
- Compulsive changers
- Cars that spend a lot of time being repaired
- Anyone who does not cherish the value of time

Consider this a list of time bandits, and recognize that each one steals one of the most precious assets in your life—one that can never be returned.

PART THREE
THE KEYS

12

Shortcuts to the Ninety-Minute Hour

The first shortcut to gaining extra time is a password, that is, one of several. This password to the ninety-minute hour is the word *while*. The more you comprehend this basic concept, the more ninety-minute hours you'll enjoy.

THE IMPORTANCE OF THE WORD *WHILE*

The unconscious but consistent question to ask is "What else can I be doing *while* I am doing this?" Of course, many times the answer will be "nothing." But this book is about all those other times when the answer allows you to complete one task while undertaking another.

I am not suggesting that you tackle two things requiring the same type of concentration. I am suggesting that you simultaneously participate in two things that don't tap the same portions of your mind.

Listening to an audiotape requires one kind of concentration. Driving requires a totally different kind. So listening while driving is not a problem. Sitting in an airplane seat calls for no concentration at all. So it's a zero-stress activity to use an Airfone while flying as a passenger. Doing any productive activity while waiting not only is stress free but often frees you from stress. Hearing a subliminal tape while writing something does not tax your mind or your emotions.

Never do I attempt anything as stressful as attending a meeting while writing a memo. Extra time is rarely worth that kind of stress. Stress is too expensive a price to pay. In contrast, delegation, if done right, is a completely stress-free way of attending to one task while another is being handled—and it doesn't even matter if both tasks require the same skills. As long as you've got someone fixing the main course and side dishes, you can focus on fixing the dessert, then enjoying the whole meal.

A METHOD FOR EASILY DOING TWO THINGS AT THE SAME TIME

Take the brief time necessary to make four lists. List 1 should contain all the personal and business chores that take up your time during a typical week. List 2 should be those chores that demand your full personal attention.

List 3 should include those chores that can be combined with other chores. List 4 should comprise those chores that can be delegated in some way.

Memorize Lists 3 and 4. Live by them. Let them dictate the tone of your business life and your personal life. Add to the lists whenever you can.

Don't cut yourself short by the limits of your imagination. If exercise is on your List 1, remember that you can put it on List 3 if you practice the many isometric exercises that can be accomplished while sitting, standing, or waiting; trade in the aerobics class for the benefits you get while exercise-walking (using it to commute if you possibly can); and substitute stair ascents and descents for elevator rides.

Stretch your imagination. While you are going shopping for someone's birthday gift, you can handle all the Christmas shopping. You can avoid holiday rushes and get in on sales with this extremely efficient timesaving measure.

Put rhythm into your work. People are known to work best at varying speeds and rhythms. There is no one right speed or rhythm for the human race, but there is one for you. The sooner you learn it, the more efficiently you'll be able to work.

"KNOWLEDGE" IS A PASSWORD TO SAVING TIME

For this reason, self-knowledge is another key to the ninety-minute hour, and *knowledge* is another shortcut. Some people function best early in the morning; others are better late in the afternoon; still others are at their peak at night. Determine which time is best for you, then set about accomplishing your personal and business chores while you are functioning at your best.

Knowledge of your own work, your own goals, and especially your best skills is one more key to the ninety-minute hour. This knowledge will enable you to fit the right tools to your work. Wrong tools take up inordinate amounts of time. Right tools do just the opposite. Tools are not better just because they are bigger. They are better if they do the job required with a minimum of effort, with a minimum of complexity, with a minimum of power.

Working at a job that does not capitalize on your best skills is theft of both your time and the time of the people who pay your wages. Appropriate application of skills is always a timesaver in the long run. It is an automatic hedge against boredom and burnout. Knowledge will connect you with those skills.

EFFECTIVE IS A PASSWORD; *EFFICIENT* IS NOT

Another crucial timesaving concept is the immense difference between being efficient and being effective. The key for you is to concentrate on being *effective.* Once you have become effective, then you should concentrate on efficiency, but not before. As famed economist-author Peter Drucker points out, "Efficiency is a minimum condition for survival after success has been achieved. Efficiency is concerned with doing things right. Effectiveness is doing the right things." Now that you have the knowledge to access your whole mind efficiently, the increase in your effectiveness should be noticeable to others and a source of pride to yourself.

While creating ninety-minute hours for yourself, always keep in mind the most heinous of all time bandits—nonproductivity. I don't have to tell you that productivity equates with the valuable use of time. But it appears that many people aren't yet aware that there is nothing less productive than the idle time of expensive capital equipment, unless it is the wasted time of highly paid people. The more you can recognize nonproductivity, the more time you can save—for yourself and your company.

A good method for maximizing your productivity is to ask regularly, "What is my business these days?" That helps you answer the questions "What business activities should I focus more on?" and "What business activities can I devote less time to?"

McDonald's once asked themselves this question and came up with the wrong answer: They thought they were in the hamburger-making and -selling business. In truth, that wasn't their business but their prime activity. Once they gained the knowledge that their prime business was to provide basic, ready-to-eat meals in a timely, clean, and efficient fashion, they decided to introduce the fast-food breakfasts that now account for up to 40 percent of the revenue of some McDonald's outlets. They introduced salads, which attracted millions of health-conscious people to opt for a fast-food meal. Earlier McDonald's had been efficient. Now they became efficient and effective, a far more profitable combination.

You'd think that a successful and enormous operation such as McDonald's wouldn't give any wrong answers to basic questions about their own operations. But keep in mind that for new ventures as well as for existing ventures, lack of market focus is one of the most typical shortcomings. It wastes energy. It cuts down on productivity. It wastes time.

Knowledge of what ought to be your prime market

focus will save your energy, increase your productivity, and maximize your time.

CHANGE AND THE NINETY-MINUTE HOUR

One more area where astute knowledge is required if you're to enjoy the ninety-minute hour: *change.* You may have made your four lists, learned what you can be accomplishing while you are accomplishing something else, delegated all that can be responsibly delegated, harnessed your unconscious power, tuned in to your optimum working hours, and focused sharply on your market. Then what happens? *Change* happens. It always does.

The key to the effective use of your time will be your recognition of the change, then your ability to adapt your work style to meet it. I cannot guarantee that you will have time for a life filled with ninety-minute hours, but I can guarantee that you'll have to cope with change.

This will entail changing your focus, your manner of delegating, the people to whom you delegate, what you delegate, how you spend your time, and how you can most effectively utilize your time.

The bad news is that regardless of your effectiveness and efficiency, you'll have to undergo some changes sooner or later. The good news is that if you have knowledge of the changes that occur in your business and personal life, you'll be able to maintain your ninety-minute hours. Be sure that the changes do not cause stress but that they provide you with that lovely balance of work, freedom, leisure, and joy.

TWO KINDS OF TIME

A final item of knowledge that you probably already know deserves a mention because there are a few people who don't have this knowledge. The item concerns time, naturally. It differentiates between standard time and quality time. It is the quality time that you are seeking. You seek it in your professional life to maximize your talents and accomplishments. You seek it in your personal life to maximize your leisure and your joy.

I can make an outline for a book until I drop, or my word processor quits because of overwork. That is work devoted to my prime focus—book writing. And it is time spent on my book. But is it quality time? To my publisher, perhaps. But to me? Not really. Quality time, in my case, involves only the writing of the book. The research and outline are necessary, to be sure. But in my literary life time spent in other than pure writing is less than quality time to me.

My wife and I have always made the distinction between standard time and quality time. As married folks, we naturally spend quite a bit of time together, especially since I've been working at home. When we move the furniture around a room to give it a different look, that takes time, but it is hardly quality time. When we take a long drive, talking together, that is quality time.

One of my clients runs a furniture store. He can spend time selling his furniture. He's great at it, and he enjoys doing it. But that is standard time—for a boss. He can also spend time training his salespeople to sell as effectively as he does. That is quality time for a boss. And that's the kind of time he should—and does—seek.

Be honest with yourself about time devoted to the

attainment of your goals, time that is valuably spent but that might be better utilized doing something else.

How can you tell when you are spending quality time? You have enough knowledge of your business to know when you've used your precious time as efficiently, effectively, and in as excellence-oriented a way as possible. You only have that feeling when you are engaged in quality time. Only when you are hitting your potential do you know you are putting in quality time. A side benefit is that putting in high-quality time rewards you both emotionally and financially.

The key is that you are focusing on the intelligent expenditure of time, not on making money. If you spend the time properly, you'll earn the proper rewards. That's why so many workaholics maintain that sorry state of affairs for themselves. They are more attuned to time spent working than quality time spent working. Not only do they deprive their lives (and their families) of balance, but they shortchange their businesses. Don't ever fall into the trap of thinking that time spent working is automatically time spent valuably. If it is not quality time and quality work, you are wasting time.

Workaholics are generally well-meaning folks engaged in a love affair with work, although it is usually more of an obsession. Although they, of all people, need the most possible ninety-minute hours, they are generally too entangled with the details of work to become actively involved in streamlining their efforts. They usually lack the talent to delegate, and they aren't oriented to achieving one thing while achieving another. Instead, they are oriented to hard work and that's all. That is not enough.

The idea is to work smart, not work hard. If you do work smart, the end result will be more time. I hope you spend it smart rather than merely spend it. Luckily, age will be of great aid in your desire to work smart and spend time smart. Age will give you the benefit of enough mistakes that lead to wisdom. Age will teach you to build a bit of free time into each day. Youth can learn that with a teacher. But age helps you learn it through experience. Use that free time to tackle emergencies, to relax, to attend to unforeseen opportunities, to recover from a bout of hard work, to plan, to investigate new ideas, and to wrap up the inevitable loose ends that pop up in everyone's life.

If you wish to avoid stress, building free time into each day is a solid-gold idea. Ulcer doctors might go bankrupt if the idea were made into a law. People who plan their schedules so tightly that there is no free time have to feel stressed out when their plans don't turn out exactly as they hoped. Aim at gaining ninety-minute hours, but be sure to luxuriate in a thirty-minute hour now and then— on a regular basis.

During your free time, you'll see that ideas form in your mind, ideas that might never have been born if you were busy attending to planned details. You don't have to set aside an hour a day for "genius ideas," but if you do, don't be surprised when you do come up with more than a few genius ideas. It's much easier to come up with breakthrough thoughts when you're not trying to think them.

A BIT OF INSIGHT INTO YOUR INSTINCTS

Allow me to give you an insight into yourself: You already know all you have to know about time and its proper utilization.

Kids come into the world with an amazingly good

sense of how they ought to be spending their time. They sleep when they ought to. They eat when they ought to. They cry when they ought to. They play when they ought to. And they crave love when they ought to. As infants, they are the very essence of the ninety-minute hour.

They are seeing while they are hearing. They are learning while they are playing. They are experiencing love while they are being fed. They are masters at intelligent time utilization. Incredibly, they practice all this ninety-minute-hour behavior without the benefit of planning. The credit goes to their instincts, to their genes.

Then what happens? They lose the instincts. They become conditioned to do one thing at a time, to concentrate on what others think they should concentrate on. Although no fault belongs to their parents, grandparents, child-care figures, or teachers, these people condition them to spend their time differently—sometimes to the kids' benefit, ofttimes to their detriment.

This is not to say that we can learn about ninety-minute hours from children. But it is to suggest that you knew at the beginning of your life what I've been getting at in this book. Then you unlearned it. Can you return to that blessed state again? I doubt it. But you certainly can try.

DON'T GET THE WRONG IDEA FROM THIS BOOK

Some people, especially those who have been frittering away their time, might pick up a book that suggests you can increase your time by as much as 50 percent and get carried away with the ideas. The problem is, however, that they go to extremes. Don't become so overorganized that you lack flexibility, that opportunities pass you by because your plans were too firm. Don't become a time-management junkie and spend so much time organizing your time that you have little time left over for accomplishing. If you spend more than five minutes a day organizing your next day, this indicates a problem.

Simply list your tasks as they come up; then, just before leaving your workday behind, prioritize the tasks by numbering them. That's really all there is to it. If you have to spend a lot of time transferring tasks from one book to another, or from one calendar to another, something is wrong. The practice of saving time should save a lot of time for you, not take a lot of time.

Don't become a compulsive clock-watcher. If you do, you will begin to consider time your enemy, not your friend. As your enemy, time can cause stress, along with many of its associated body ills. If you treat time as your buddy, you'll have a good relationship with it and neither of you will mistreat each other.

SAVING MAJOR CHUNKS OF TIME

You can get and give far more if you spend less time driving and more time doing. I, who love driving, can think of thirty things I'd rather do than commute. And you can probably think of twice as many.

Now that you've learned how to make the most of your travel time, let me to introduce a radical ninety-minute-hour thought to eliminate that travel time: *Move closer to where you do what you do.* Spend less time on the road.

If it takes forty-five minutes right now for you to get

from where you live to where you work, and you'll make the commute for the next thirty years, you're robbing yourself of 11,250 hours of lifetime. That's one year and 103 days.

Consider moving. Cut that commute down to fifteen minutes. Then your travel time results in a loss of only 3,750 hours, only 156 days of lifetime. You waste less than a year on the road with this radical move.

Even more radical is the idea of *working from home.* That cuts the lost time on the road down to zero. If you're a professional person, perhaps you can do it. Maybe you can start free-lancing or consulting or doing large things from a small base at home. Perhaps you can set up your own electronic cottage and handle your work by computer.

It's usually quite difficult to move closer to work, let alone work from home. But any book that suggests methods by which you can add more to your lifetime would be remiss if it neglected to plant this thought in your mind. And you'd be remiss to dismiss it without doing some serious mulling first.

In the category of giant lumps of lifetime comes the idea of *possibly* reducing your sleep time by thirty minutes nightly. See if a bit less sleep feels comfortable. After all, this small cut will give you 182 hours (that's over 7 days) of freedom each year. Earlier, I suggested that you not cut out any sleep. Now, I suggest that you experience the feeling of cutting half an hour each night. You may appreciate the time more than the sleep.

Are there other, less drastic steps you can take to free up more time for yourself? The answer is "of course." And those steps are but a page away.

13

Recognizing the Value of Time

As you learn about time, you begin to realize how little it is related to clocks and how much it is related to priorities. You start to connect time with energy as you understand that if you waste your energy attending to things better handled in other ways, you are wasting your time.

One key to time is the passwords, *while, knowledge,* and *effective.* Another key is your conscious recognition of the value of time and the prodigious amount of time wasted by ignorance of its value. The few moments you take to award your full awareness to the limited nature and limitless opportunities afforded by time will serve as your key to the ninety-minute hour.

Many people want to be millionaires, but few become millionaires. Many people want keys to extra time, but few have them. Your recognition of time's importance is a key for you.

Now that you can see how increased efficiency can mean increased effectiveness, you're likely to increase your own. You have both the idea and the tools. You can add more time to your life. You really can add as much as 50 percent more—and possibly more than that.

Starting on a conscious level, it is very important that you know your commitment to time. That commitment is more easily and certainly kept if your associates, friends, and relatives are equally aware of your recognition of time's inherent value—beyond anything with price tags.

By going public with your new devotion to time, you will automatically begin to save quantities of time because associates, friends, and relatives will call and drop in on you less frequently, unless their contact is worth your time. They'll have increased respect for your time because you'll be demonstrating increased respect for your time.

Your phone calls will become shorter—when you want them to be. You'll see a considerable amount of work, including detail work, that ought to be tended to by someone other than you—and you'll do something to eliminate its consumption of your time. You'll look into getting more information from newsletters. You'll give a shot to subliminal suggestion—and you'll be secretly surprised at how well it works for you. Sounds like you're well on your way to the ninety-minute hour. But wait . . .

WHAT ARE YOU GOING TO DO ABOUT THE OBSTACLES TO YOUR TIME?

As increasingly available as they're becoming, ninety-minute hours don't grow on trees. You'll have to face barriers to your time extension. Some are of the obvious variety, others more covert.

- The "tyranny of the urgent" is how one business management consultant describes one of those barriers. You're moving along at hyperefficient speed, when all of a sudden an emergency looms. What does this do to your momentum? Depends on your priorities. Emergencies are far more easily handled if you know ahead of time who will handle which kind of emergency, and what constitutes a real-life emergency. Of course, urgent situations often require your hands-on help right then and there. But not always. You've got to discriminate between the elements of urgency and the elements of paranoia.

- Slavery to the telephone is another time bandit. I speak with the enlightenment of a onetime slave. You can save tremendous amounts of time by selectively fielding your calls—taking those that honestly require your time and postponing those that don't. This can be accomplished through a secretary who screens your calls, an answering service that relates them to you on request, and/or an answering device that allows you to hear the caller without having to respond—except to those to which you need to. You'd be depressed to learn how much time is lost just in unnecessary phone conversations. And it's not only the telephone time that is lost. It's also the time required to get back up to speed on whatever activity was interrupted by the call.

- Many people are time bandits. These are the people who have not yet granted time the same recognition you have. They take your time in person, on the phone, by the written word, in the hall, and even in the john. As time bandits, they steal a possession more dear to you than most of what you own. In all too many cases the time bandit is your boss. I know of three time bandits in my life. When they call, which is more and more infrequently, we talk for less than one minute. When they call during leisure time, our conversations go on and on. It is then that I realize how much time they unconsciously, but oh-so-greedily, consume.

- All time is not created equal. So use your time with the greatest fervor during the first two and a half hours you work. Studies show that managers are most productive in the morning, specifically, in their first two and a half hours on the job. Handle your most substantial priorities during that time: your most important decisions, your key telephone calls. Talk with the people who can best advance your mission. Limit your incoming phone calls to the most pressing issues. And attend meetings only if you must. Don't forget that other ninety-minute-hour seekers will be doing the same. So the demands on your time will be the greatest during those morning hours. You may hear of business success stories built on people burning the midnight oil. Maybe so. But more such tales of success are set between the hours of 9:00 and 11:30 A.M., Monday through Friday.

- Organize your work space so that the most critical information is accessible from a central file cabinet or computer terminal. The art and science of efficiently designed office environments develops regularly. An annual show, "Workspace," is held to display advances in the industry. Efficiency experts tell us that the American executive wastes an average of three hours a week searching for lost or misplaced documents. That's nearly a full week (six and a half days) every year. Simply by responsible filing and space planning, you can gain the equivalent of a week's vacation. I recommend the beaches and Mayan ruins south of Akumal in the Yucatán region of Mexico. Bora Bora is nice, too. And don't miss Venice, whatever you do. Start your holiday time by keeping current information within arm's reach. Discard obsolete materials regularly, thrilling in the freedom from their oppression. The fewer materials there are around, the easier it is to find things.

- Embrace the new timesaving technologies. Even if they seem silly to you at first, they'll soon win a place in your heart. Example: My wife and I have installed a device called an instant hot in our kitchen. It constantly maintains a supply of boiling water. When my wife wants a cup of tea, she merely turns the spigot. Same for my morning coffee— freshly ground, but instantly prepared. Ninety- minute hours often come in small groups of minutes. This is one example of them. It takes about five minutes to bring a teapot to boil. If you drink only two cups of tea or coffee per day, that's a timesavings of ten minutes. If the household consumes six cups of hot tea or coffee daily, that's a savings of thirty minutes. During the course of a week, the savings goes up to three hundred fifty minutes. In one month, that comes to a savings of fifteen hours. You can gain the equivalent of more than half an extra day per month with one simple appliance that costs less than one hundred dollars. Of all the fancy electronic doodads in our house, few have saved us as much time as that instant hot.

- Be aware of constantly introduced products developed to save your time. A new electronic voice message recorder lets you leave messages for the people you live with without spending the time to write them down. A new plant waterer handles the watering of your plants for you (and costs only $8.95). Available now are light bulbs that last eighty times longer than standard light bulbs. That means you can save the time it takes to change a light bulb, and you can save it seventy-nine times. And that's just for one light bulb. These bulbs, priced at about five dollars each, can save you a large amount of time. A new device lets you know when your mail is delivered, saving you the time spent checking the mailbox. A new compact disk player offers six straight hours of music, all selected by you. That's quite a bit of time saved changing CDs. These are but a sampling of the timesavers that are being introduced regularly. Keep an eye open for them. They are even more tools for ninety-minute hours.

- Avoid clutter mentality. Clear your desk daily to keep your work space, your life, and your mentality free from clutter. Clutter can overwhelm some individuals, giving their work a plodding rhythm. My desk may be cluttered while I work, but it is spotless when I'm through. Everything gets dealt with or put into a file.

- I have three sets of files: current, not-so-current-but-alive-at-any-moment, and hold-a-bit-longer-just-in-case. During free but unplanned moments, I move the current (in my file right at my fingertips) over to the not-so-current (in a cabinet across the room). I move the not-so-current to the hold-a-bit-longer (in a storage shed outside my house). And about once each year or two, I transfer the hold-a-bit-longer to the garbage. In the fifteen years I've been using this system, it has yet to come back to haunt me. When I transfer my old files to the garbage, I feel 200 pounds lighter. And I weigh only 160.

- Create no undue pressure for yourself. There's enough pressure out there without your adding more. Daniel Stamp, president of Priority Management Systems in Dallas, says, "The average person has thirty-six hours of work on his desk at any given time. Far more than he can ever hope to accomplish in a workday. Having these documents in full view only serves as a reminder of what the manager cannot accomplish, creating undue pressure and prompting him to tackle projects randomly rather than prioritizing his work load." Stamp reminds us all that doing the job right is fine, but doing the right job at the right time is even better. I certainly wouldn't argue. And I enthusiastically second his suggestion about being ultracareful not to create your own stress.

- Frequently, there will be conflicting demands on your time. If ever you need a ninety-minute hour, that's when you need it. But it's not there then. Listening to an audiotape isn't much help when two fires are roaring, threatening to spread out of control; when two deadlines are approaching, and there's only one of you. Or is there? Delegation is one solution to conflicting time demands. Rescheduling is another. Prioritizing with skill is a third.

- Richard Nelson Bolles, author of the immensely popular and effective book on the almighty job hunt, *What Color Is Your Parachute?* (Berkeley: Ten-Speed Press, updated annually) says this about prioritizing: "I think almost everybody today has some problem with time. They are never, ever going to get done all that they want to do and therefore they have to establish priorities. They have to get a vision in their head of what is most important to address." It's doubtful if all the demands on your time are of equal importance to your business or life on earth. Simply decide which most merits your time and delegate, reschedule, or even cancel the other. If conflicting time demands create stress for you, try to foresee them better. Get that vision in your head of which is most important to address. This will eliminate the stress, which Bolles calls "the pain of time," if not the conflicting demands themselves.

- As more companies are discovering the power of mail, you are being besieged with more and more direct mail. It's hard for me to knock it since I earn part of my living writing it. But it is easy for me to warn you to be choosy in what you read. Don't make the mistake of classifying all business mail as junk mail, because some of it is anything but junky. The idea is to have someone screen for you or to become a talented screener yourself. Not easy, but it comes with practice. I know it's tough to toss away an envelope that announces you've won a round-the-world trip, a Rolls-Royce, $10 million, and a castle in Spain. But if you're going to be efficient about time, you've got to be tough about the mail, castle or no castle.

- A downside of the Information Age is too much information. You ask for background material and someone hands you seven cartons of facts. Your job here is to convince the people who gave you the cartons that it is their job to winnow the information down to only the most pertinent. Ask them to put that in a folder, then summarize the contents of the folder on a single page. Now we're talking ninety-minute hours.

SLOW IS BEAUTIFUL

With all my paeans to the glories of time, I hope you never shortchange yourself trying to save a couple of minutes. I don't mean just avoiding getting caught going over the speed limit. I also mean not missing out on the quality of life's offerings simply to save time.

Saving time is nice, but it surely should not be the prime goal of anyone's life. For instance, speed-reading can save you a lot of time. But don't speed-read everything. Woody Allen once remarked in a movie role, when referring to his having taken speed-reading, "I read *War and Peace* in fifteen minutes. It's about Russia." My own daughter, an ace speed-reader while in her midteens, read *The Godfather* in under two hours. My wife and I were duly impressed until we saw her reading the book a second time, then a third time, then a fourth time. Eventually, she spent as much time with Mario Puzo as we did.

Speed-readers don't actually comprehend much more than the gist of a text unless it deals with familiar terrain. You certainly wouldn't want your surgeon speed-reading the new findings about your impending surgery. You wouldn't want your lawyer speed-reading a contract that will commit you to an expensive purchase.

Evelyn Wood herself, who opened her Reading Dynamics Institute in 1959, says, "A person who speed-reads and does it correctly can read with better comprehension. I'm sure of it." But most people who take her course are professionals trying to clear the paperwork from their desks and get through their journals faster.

If you have the time and opportunity to read a truly good book, I recommend non-speed-reading every time. If you are presented with the opportunity to enjoy a good meal, even a medium meal, I heartily recommend savoring it slowly. I'm about the slowest eater of any ninety-minute-hour practitioner I know. There is a great joy, especially to speed freaks like you and me, in doing some things slowly.

BUT SLOW ISN'T ALWAYS BEAUTIFUL

In the world of business, slowness is the equivalent of sinfulness. All too frequently, the big problem in business is lack of time. Well, Mr. and Ms. Businessperson, time extension is the solution. The ninety-minute hour is the solution—as many ninety-minute hours as possible. With lack of speed presenting so major and common a problem, the new technological and psychological findings presented in these pages are very welcome.

But far too few companies have the time to spend learning how to save time and create time. These companies should assess their managers, then provide them with the tools and the training to benefit from as many ninety-minute hours as possible. The companies will benefit. The managers will benefit. The customers will benefit. Business in general will benefit.

A "to do" list is a list of things to do with a priority attached to each task, as indicated by simple numbering. Almost every conscientious person knows of such lists. Most have made out such lists. But not many make out "to do" lists on a daily basis.

If people have a particularly heavy load, they may make a "to do" list. Sometimes, with a myriad of deadlines staring them in the face, they'll make up a list. But the key is to do it daily. Daily time planning is one of the things that separates the leaders from the followers. The leaders know that they are in charge of time. The followers feel that they are the victims of time.

Those who keep "to do" lists daily derive great satisfaction from crossing off the tasks as they are accomplished. They almost always cross off every task they've listed. They never lose their lists, as some would-be timesavers do. They wouldn't dream of keeping their lists in their heads. They know doing so would not only clutter their brains but would also clog up the pathways through which one thought collides with another to create a gem of an idea.

Keeping your list on a daily basis is just one-third of your job. The second third is prioritizing the list, separating the musts from the maybes. The final third is completing everything on your list.

I hope you use these timesaving hints, these keys to the ninety-minute hour in both your business and your personal life. A Stanford University study of the super-healthy—five hundred individuals who never complained of headaches, backaches, other pains, colds, or the plethora of problems that afflict humanity—revealed that among the few things these people had in common was that they were well organized in their business and personal lives.

Other things they had in common, as long as we're on the topic of maintaining health, were

- They tended to act on their feelings.

- They had a creative outlet somewhere. Even the bank president did wood carving as a hobby.

- They exercised regularly to the full extent of their limitations. The key words are *regularly,* which does not mean daily, and *limitations,* which does not imply four-minute miles or four hundred push-ups.

I only tell you this to show the importance of time organization to good health. And as long as you're going to have all that extra time, please don't spend it feeling under the weather.

TIPS FOR GAINING TIDBITS OF NINETY-MINUTE HOURS

Here's a collection of ways to save extra minutes here and there at home, in the office, and in between. Each time-saving method is small when compared with delegating work or gaining an education while listening to music or driving to and from the office. But these extra minutes will combine to make extra hours.

This brief list is far from complete. It's up to your unconscious to add to it. These are simply mind stimulators, ideas that are designed to set into motion a series of other ideas—some obvious, others less so.

1. Clean up as you go along so that you don't have to retrace your steps to do the cleaning.
2. Set your table with just-washed dishes. Unless there is a crying need for them to go from the sink or dishwasher into a cabinet, put them right back on the table.
3. While you're cooking, put items into the dishwasher. Don't separate the cooking from the cleaning. You'll save time and have a neater kitchen.
4. Walk faster. It doesn't take much to quicken your pace a smidgen, and doing so will probably be good for your body. If you walk a considerable distance each day, speeding up a bit can save you a significant amount of time on an annual basis.

5. Plan your meals and your attire ahead of time. Everyone knows that planning under pressure takes longer and is more inefficient. Anyway, how can you trust your sartorial instincts by dawn's early light?

6. Cook double or triple the amount of food you need and freeze what you won't eat at the moment. Done properly, this can be a fairly generous saver of time. Freeze the finished dishes in meal-sized portions.

7. Use a microwave oven to save a significant amount of time. (I'm only including so obvious a timesaver because I just discovered it.)

8. Invest in a large freezer so that you can enjoy the time- (and money-) saving benefits of shopping in bulk.

9. Pick up your perishables on the way home. Don't make a special trip to get them. Travel home from work on a route that takes you past a grocery store.

10. Take advantage of the new delivery services that are popping up all over the place. In San Francisco a company will take your grocery order by phone, then deliver it to your house. In Tokyo a company will take your book order by phone, then deliver the books you want to where you live. In other places companies will supply you with a directory of videos for rent, then, after receiving a phone call from you, will deliver and later pick up your rental videotape. Many cities now have mobile mechanics who will come to your house to do all the mechanical or body work on your car so that you don't have to spend your

valuable time dropping off the car, doing without the car, then picking up the car. Because of the increasingly precious nature of time, home and office delivery services are among America's fastest-growing businesses. Meal delivery services are burgeoning as well. Busy people *need* those kinds of services.

11. Any time you handle any piece of paper, limit your contact with it to one time. If it requires a response, respond now—not later. If it must be filed, file it now. If it might be better off in your wastebasket, toss it now.

12. Learn how to say no. This is not an easy word to say, and you'll have to say it to some of the nicest people in the world. Don't say it whenever you can avoid it, but do say it when that inner voice of yours—the one that seems a mite louder now that you are a devotee of the ninety-minute hour—tells you to. When you do say no, say it immediately rather than leaving people hanging and wasting their time. Be nice, direct, compassionate, and even loving. But be negative when you know you should be.

13. Respect your instincts and your mood. If you're not in the mood to do something, do something else, saving the original task for a time when you are more likely to tackle it with aplomb. Otherwise, you'll end up taking more time with the project than it is worth because, deep down, you don't want to do it.

14. Make a public commitment. If you want to complete a task by a certain hour, go public with your pronouncement. "I'll have that report to you by

three o'clock" leaves you with little choice. Either you are an efficient person who keeps her or his word, or you're a blatant liar, not to mention a procrastinator. A simple choice for most of us to make.

15. Take breaks. If you work straight through without any break at all, chances are your work will lag. But if you take brief breaks, you'll be able to keep up a healthy and rapid pace. Your work will also be more accurate, according to the experts.

16. Learn to tolerate your faults in an effort to overcome perfectionism. You'll have a far better time of almost everything in life if you strive for excellence rather than perfection. This trait will also help you delegate with more success.

17. Get things done right now. It is estimated that a minimum of 80 percent of the business coming across the desk of an efficient executive gets handled immediately, through either personal action or delegation. Nothing is uglier to such a person than an "in" basket piled high.

18. Since neatness leads to organization and organization leads to time efficiency, the time experts recommend forcing neatness on yourself. If you won't be using an item during the next two hours, put it away in its proper place.

19. When you've got a mountain of tasks or details, take heart that you'll be dealing with them one at a time. This simple mind-set can help keep you well organized because one of the main reasons some people can't seem to organize anything is that they have a deep-seated fear of having to organize *every*thing.

20. When negotiating, realize that the shortest route to an agreement isn't necessarily the straightest. Be prepared to manage the negotiations through the side streets that lead to the final destination. It takes 90 percent of the total discussion time to resolve 10 percent of the issues—and the final 10 percent of the time to resolve the other 90 percent. By recognizing this negotiation idiosyncrasy, you can save a great deal of valuable time.

STOP ALL THAT TIMESAVING

One of the best things about saving time is that it provides you with enough time to stop accomplishing. The cessation of energy expenditure helps contribute to your body rhythm, which works in ninety-minute cycles. That suggests that you should take a break of a few minutes during each cycle.

The break can come in any form: a phone call, a walk, a different activity, a conversation, a few minutes of self-hypnosis or meditation, a session at the watercooler. Your major benefit from the break will be the change in activity. So you can gain from both rest breaks and work breaks. Breaks contribute to a larger, longer, deeper rhythm—the rhythm of balance.

KEEP AN EAGLE EYE OUT FOR TIMESAVING SNIPPETS

As you gain appreciation of time by gaining awareness of wasted time, you'll pick up snippets of timesaving information from sources as diverse as movies, newspapers, and friends. Here are a few samples.

Timesaving snippet: You can clean while you polish if you use club soda or rubbing alcohol. Both do both jobs at the same time, completely eliminating one task.

Timesaving snippet: You can become involved with some of the best seminars in the nation without leaving your home. Many top seminars are available on audiotape. Unless it's about something that requires physical presence, a seminar on tape is a way to gain information while saving time. Use your time compression machine.

Timesaving snippet: Eat a light lunch so that digestive action doesn't slow you down in the afternoon. You can eat in your office and accomplish work while you eat, but I don't recommend carrying the ninety-minute hour to such extremes. Working during lunchtime is unquestionably extreme. Extremely efficient and extremely inhuman.

WASTED EFFORT IS WASTED TIME

As there are bandits who steal time, there are misguided efforts that rob you of time. Here are ten of the more everyday kinds of efforts that are misguided if you don't want to waste your time:

1. Reading unnecessary material. It's in your magazine, so you should read it. Wrong. Eliminate nonessential stuff.
2. Doing instead of delegating. It's hard to believe someone else can do it as well as you, but they can.
3. Daydreaming. Not an unhealthy activity unless

you get too caught up with it. Daily daydreaming sounds like overdoing it.

4. Worrying. We all know the dazzling benefits of worry: success, health, accomplishment, happiness. I'm just kidding; worrying is no way to save time, only to waste time.

5. Rewarding yourself before you deserve it. Everyone deserves a reward for a job well done—after it is done.

6. Being social. A definite change from working during work is socializing during work. It's a time-waster.

7. Getting too involved. You'd be delighted at how some of those details that entrap you are able to handle themselves.

8. Skipping out. I know it makes you feel good to get away from work, but timing is everything; pick yours with care.

9. Doing something else instead. You've got to write the report, but you want to reorganize those files. Bad idea to deal with the files now. Deal with them later.

10. Not realizing you're wasting time. That's a key to the ninety-minute-hour kingdom. Recognize time's value.

One key to the ninety-minute hour is knowing the passwords. Another key is recognizing the value of time. The third key is your willingness to devote a few minutes now to putting some ninety-minute hours into your life.

14

Countdown to the Ninety-Minute Hour

Let's swing into the world of the ninety-minute hour right this moment with a short chapter and a short request of you. It's an offer you'd be nuts to refuse because it will give you more time, more life, and more freedom at a cost of only a few minutes of your time. I know them to be valuable minutes, so I'll get right down to the process of putting more minutes into your hours.

BEFORE THE COUNTDOWN, THE COMMITMENT . . .

The benefits of one life filled with more than one lifetime begins with you deciding in your own mind, right now, that you sincerely do want to begin to gain more time for yourself. That decision implies a commitment to yourself. Please don't take it lightly, because that would mean you take yourself too lightly. Not making the commitment is a lot better than making it and not keeping it. Making it and keeping it is best of all.

STEP 1 OF THE COUNTDOWN: THE FIVE LISTS

A lifetime of ninety-minute hours begins with only a three-step countdown. The first step is easy: sit down with a pen and a sheet of paper and make the four lists I suggested in Chapter Twelve, plus one more list that I saved for now.

List 1 includes the business and personal chores you do in the course of a month. It ought to take about five or ten minutes to compile this list.

List 2 includes the chores that require your full attention and only yours.

List 3 includes those chores that can be combined with other chores and opportunities. This will include both the chores you currently do and opportunities to learn, grow, improve, earn, excel, and succeed.

List 4 includes those chores that can be delegated to someone else. This is where you can get involved in some serious timesaving. If it costs money to delegate, remember that your time is more valuable than money.

List 5 includes those chores that you can drop. There's a good chance that you're doing a few things that really don't have to be done at all. I'm not suggesting you drop activities that are enjoyable, just those that might be a tad compulsive. The idea is to streamline your life.

Living by the insights you gain from Lists 3, 4, and 5 can put you on the road to many a ninety-minute hour. Have those lists made into wallpaper and paper your home with them. Or do something else that will implant them deep in your mind. Here's a way to accomplish that.

STEP 2 OF THE COUNTDOWN: THE AFFIRMATIONS

You know from Chapter Eleven some proven techniques for accessing your unconscious mind. Here's a way to breathe life into them. Now that you know how to gain extra time for yourself, write an affirmation about what you're going to do with that new information. You can write two or even as many as five or more affirmations. But be sure that these affirmations will provide extra time for you.

It's important that you actually put your affirmations into writing so that you can see and read them. Writing your affirmations is making your commitment to the deepest, most powerful part of yourself. These affirmations will change your lists from timesaving theory into timesaving action.

Now comes the fun part.

STEP 3 OF THE COUNTDOWN: THE FREE TIME

Make a list of what you'll do with the free time you'll be creating for yourself. Know ahead of time what you'll do with all that extra time, or you won't appreciate it or use it wisely. You can use it for work, for recreation, for rest, for creation, for virtually anything you want. Still, it's important that you take the three minutes necessary to put in writing how you'll use your newfound free time. Otherwise, free time can bore you or get you into trouble.

Like money and power, free time is an innocent force that can be used, misused, or abused. It is not automatically wonderful. You've got to know how to use it with intelligence. If you don't, you won't appreciate the free time. If you do, you'll consider it a blessing.

Ninety-minute hours are intriguing to consider. They're shining in their promise. And they are available to you now. No waiting.

For some people, ninety-minute hours will remain a fascinating theory, and these people will continue on as their ancestors did, acting on the same information base, using the same tools, limiting their time for accomplishment and achievement, as did their forebears. But many people will recognize the simplicity of making ninety-minute hours a glorious reality for their lives. The benefits of extra time will be too delightful for them to deal with ninety-minute hours on an intellectual level only. They'll know that gaining extra time calls for action. And they'll be willing, even eager to take that action.

By incorporating into your everyday behavior the tools and keys of the ninety-minute hour, you will gain the priceless gift of extra time, and its benefits will be yours to do with as you wish. Ninety-minute hours offer you the equivalent of a longer life span. They offer you the opportunity to gain the information to fill that increased life span with achievements. They provide abundant time for leisure and the pursuit of pure joy. Ninety-minute hours present freedom from stress. They offer more time without more pressure.

The idea is, naturally, for you to enjoy a lifetime of ninety-minute hours. The idea will have achieved lift-off if you see time in a new way and act in a manner befitting its value.

There have never before been so many opportunities for you to save so much time, to add so many minutes to so many hours of your life. As you comprehend the alliance between your time and your life, you will know how to expand both.

It's about time.

Products, Services, and Information for the Ninety-Minute Hour

BOOKS

CULP, STEPHANIE. *Getting Organized.* Cincinnati: Writer's Digest Books, 1986.

DIXON, N. F. *Subliminal Perception: The Nature of a Controversy.* London: McGraw-Hill, 1971.

DRUCKER, PETER. *The Effective Executive.* New York: Harper & Row, 1967.

FERNER, JACK D. *Successful Time Management.* New York: John Wiley & Sons, 1980.

FINK, DIANA DARLEY, ET AL. *Speed-Reading.* New York: John Wiley & Sons, 1982.

GAWAIN, SHAKTI. *Creative Visualization.* Berkeley: Whatever Publishing, 1978.

GODIN, SETH, AND CHIP CONLEY. *Business Rules of Thumb.* New York: Warner Books, 1987.

HEMPHILL, BARBARA. *Taming the Paper Tiger.* New York: Dodd, Mead, 1987.

KEY, WILSON BRYAN. *Subliminal Seduction.* New York: Prentice-Hall, 1973.

KORN, ERROL R., ET AL. *Hyper-Performance.* New York: John Wiley & Sons, 1984.

LAKEIN, ALAN. *How to Get Control of Your Time and Your Life.* New York: David McKay, 1973.

LEBOEUF, MICHAEL. *Working Smart.* New York: Warner Books, 1979.

MAYER, IRA. *The Electronic Mailbox.* New York: John Wiley & Sons, 1986.

MORRIS, FREDA. *Self-hypnosis in Two Days.* New York: E. P. Dutton, 1975.

CATALOGS

Adventures in Learning. 1260 Hornby Street, Vancouver, BC, Canada V6Z 1W2 (800-663-0222). Audiocassettes offering subliminal stimulation, hypnotic suggestion, or straight information, including *Stopping Procrastination* and *Self-Hypnosis.* Also carries a broad selection of self-improvement videotapes.

Audio-Cassette Workbook Programs. American Management Association, Extension Institute, P.O. Box 1026, Saranac Lake, NY 12983 (212-903-8040). A series of audiocassette workbook programs consisting of four to six cassettes plus a workbook, including titles such as *Successful Delegation* and *Total Time Management.*

Audio Editions. 1900 South Norfolk Street, San Mateo, CA 94403 (800-231-4261). Audiocassettes of nonfiction and fiction, including *Mackenzie—On Time* by R. Alec Mackenzie, and *No Nonsense Delegation* by Dale McConkey.

Careertrack Catalog, The. 1800 Thirty-eighth Street, Boulder, CO 80301 (800-334-1018). Audiocassettes, seminars on tape, live seminars, and on-site training, covering a multitude of career advancement topics, including *Getting Things Done.* Catalog also offers a device which plays audiocassettes in half the time without the "chipmunk" voice associated with speeded playback.

Catalog of Time-Savers, The. The Ninety-Minute Hour Institute, P.O. Box 1336, 260 Cascade Drive, Mill Valley, CA 94942 (800-748-6444). The latest products, tapes, and services for saving large and small amounts of time, the pick of the technologies.

Products for the Ninety-Minute Hour

Discoveries Through Inner Quests. Institute of Human Development, P.O. Box 1616, Ojai, CA 93030 (800-443-0100, ext. 356). Audiocassettes featuring subliminal tapes that combine guided relaxation, visualization, subliminal communication, and affirmations on various self-development topics, including *The Final Solution to Procrastination, Visualization: The Key to Realizing Your Dreams, Sleep Reduction, Manage Your Time,* and *Becoming Efficient and Organized.*

Love Tapes, The. Effective Learning Systems, Inc., 5221 Edina Industrial Blvd., Edina, MN 55435 (612-893-1680). Audiocassettes, available with standard and subliminal suggestions, covering a wide range of self-improvement topics, including *Visualization Power, Overcoming Procrastination,* and a variety of courses and meditative music backgrounds.

Mind Communications. P.O. Box 904, 1844 Porter S.W., P.O. Box 904, Grand Rapids, MI 49509 (800-237-1974). Audiocassettes featuring subliminal suggestion and numerous self-improvement topics, including *Reducing Sleep, Efficient and Organized, Time Management, Learning to Say No,* and *Do It Now.* Subliminal videotapes are also available.

Psychodynamics Research Institute. P.O. Box 875, Zephyr Cove, NV 89448 (702-588-7999). Audiocassettes incorporating subliminal stimuli and time compression, including *Procrastination,* and *Let's Get Organized.* They also offer custom-designed tapes and transmitters that allow subliminal suggestions while you listen to the radio.

Psychology Today Tapes. P.O. Box 059073, Brooklyn, NY 11205 (800-345-8112). Audiocassettes on developing your potential, improving the workplace, enriching your relationships, gaining control, expanding awareness, and therapeutic techniques.

Sharper Image, The. 650 Davis Street, San Francisco, CA 94111 (800-344-4444). A selection of high-tech, high-style, and time-extension products, including a paging system, speed-typing teacher, microcassette recorder, automatic watering device, hands-free telephone, cordless telephone, answering device, portable CB radio, even a microwave oven.

Subliminal Plus. The Randolph Tapes, Success Educational Institute International, P.O. Box 90608, 2108 Garnet Avenue, San Diego, CA 92109 (800-248-2737). Audiocassettes offering subliminal suggestions and covering many self-improvement topics, including *Stop Procrastination* and *Time Management.* Catalog also offers self-improvement videotapes.

SyberVision Catalog, The. 7133 Koll Center Parkway, Pleasanton, CA 94566 (800-777-5885). Video- and audiocassette programs oriented to self-improvement in business, sports, and general living. SyberVision uses visual images to bypass conscious thinking and directly access the unconscious. Topics include *Self-discipline* and *Achievement.*

SERVICES

HYPNOSIS

American Society of Clinical Hypnosis. 2250 East Devon Avenue, Suite 336, Des Plaines, IL 60018. Will provide data on professionals using hypnosis in your area.

OFFICE SERVICES

HQ Services & Offices. 900 Larkspur Landing Circle, Suite 201, Larkspur, CA 94939, (415-461-6744). National network of people to delegate to: typists, word processors, secretarial support. Also offer phone service, facsimile transmissions, and offices. Write or call for location nearest you.

SPACE MANAGEMENT

Workspace. 665 Chestnut Street, San Francisco, CA 94133 (415-776-2111). Annual national exhibition and conference on office environments, seminars on space management.

TIME MANAGEMENT

Reynolds Communications. 6004 Avenida Cresta, La Jolla, CA 92037 (619-459-4149). Time management for company departments.
The Ninety-Minute Hour Institute. Box 1336, 260 Cascade Drive, Mill Valley, CA 94942 (800-748-6444). Time-extension consulting services for companies and individuals.
Time Power, The Charles R. Hobbs Corporation, P.O. Box 21567, Salt Lake City, UT 84121 (800-332-9929). Conducts seminars throughout the nation.

AUDIOCASSETTE TOPICS

Here's a partial list of topics you can learn more about without devoting any extra time to doing it:

Achievement—There's more to be said about getting out of the rut and onto the winning track than words of motivation. There are basic understandings, and tapes impart them.

Assertiveness—The lack of assertiveness holds back many good people in both work and relationships. Now tapes help you strike the balance between powerful and nice.

Communication—For many people the right ideas are in their minds, but the right ways of stating them don't come easily enough to let others know of those ideas. Audiocassettes can increase your influence and persuasiveness.

Creating Teamwork—Getting the most and best from people is important. The ideas can be applied in business, at home, on camping trips, and even with teams. The expertise spawns leaders.

Getting Results—Style is nice and pleases the eye, but results is the name of most of the games. Audiocassettes give insights and techniques for producing results.

Leadership—Want to be the boss? Cassette courses teach you the major elements of leadership, instruct you on taking charge in your business and personal life, in every situation.

Power—There's at least one cassette designed to show you how to gain instant rapport, overcome your limitations, and be in command of even the smallest detail.

Self-discipline—At least one course isolates the ten essential characteristics of highly self-disciplined individuals, then teaches you how to apply them to your life.

Success—Courses on this popular goal give you information on career skills. They're about goal setting, self-esteem, shortcuts to success, how to accomplish more.

Vocabulary Building—Maybe your inability to find (or understand) the right word at the right time holds you back. Cassette courses can permanently enlarge your vocabulary.

That's just the beginning. And the industry of providing the information you want is just beginning too. I'm starting to feel like Dale Carnegie as I write so positively of the available cassette courses, so I'll just kick back and list some of the others:

Achieving Excellence
The Art of Innovation: How to Be a Change Master
Dictating Effectively
Getting Rich Your Own Way
Getting Things Done
Goals

201

How to Be a No-Limit Person
How to Deal with Difficult People
How to Listen Powerfully
How to Speak Up, Set Limits and Say No
Image and Self-projection
Making It (in Business)
The New Masters of Excellence
Political Savvy
Power Memory
Profiles of Achievement
The Psychology of High Self-esteem
The Psychology of Selling
The Secrets of Power Negotiating
Speaking to Win
Stress Management for Professionals
Successful Self-programming
Think and Grow Rich
24 Techniques for Closing the Sale
Yes! You Can Write

And just to take your mind off business for a moment . . .

Attracting More Love
The Be Happy Attitudes
Being a Happy, Effective Parent
Blood Pressure
Business Success
Concentration
Coping with the Death of a Loved One
Decision Making
Deep Relaxation
Developing Enthusiasm
Developing Your Creativity
Developing Your ESP
Effective Speaking (Without Fear)
Effective Studying and Test Taking
Energy
Exercise Motivation
Getting a Good Job
Goal Achievement
Guilt Free
Health

How to Attract Money
How to Be Happy
How to Be Positive
Improving Relationships (Old and New)
Inner Game of Selling
The Joy of Loving
The Joy of Pregnancy and Childbirth
The Joy of Sobriety
Learning Power
Loving Your Body
Managing Stress
Memory
Motivation
Overcoming Procrastination
Overcoming Worry
Pain Relief (Headaches and Other Pain)
Peak Performance in Sports
Permanent Weight Loss
PMS (Premenstrual Syndrome)
Relationship Strategies
Restful, Revitalizing Sleep
Self-confidence
Self-image for Children
Self-image Programming
Sense of Humor
Slim Image Weight Control
Smoking Control
Spiritual Healing
Staying Young
Surviving Separation or Divorce
Taking Charge of Your Life
Visualization Power
Winning

And if that wasn't New Age enough for you, there's also music created especially for meditation as an antidote for stressful times, for feelings of well-being. Examples:

An Island Called Paradise
Inner Peace
Inspiration
Music for an Inner Journey

203

Omni Suite
Serenity
The Secret Garden
Simplicity

From the cream of self-improvement best-sellers, among many, you can select:

Choosing Your Own Greatness by Dr. Wayne Dyer
Love by Dr. Leo Buscaglia
Tough Times Never Last but Tough People Do by Dr. R. H. Schuller

Getting back to business, and homing in on details, consider

Communications Skills for Secretaries
Computer Fundamentals for Managers
Constructive Discipline for Supervisors
Creative Problem Solving
Dealing with Difficult Behavior
The Effective Executive
Executive Writing, Speaking and Listening Skills
Finance and Accounting for Nonfinancial Managers
Fundamentals of Budgeting
Get That Job!
Guerrilla Marketing
How to Be an Effective Middle Manager
How to Beat Your Quota
How to Improve Customer Service
How to Interpret Financial Statements
How to Interview Properly
How to Make Money in Wall Street
How to Market by Telephone
How to Negotiate
How to Speak Persuasively
The Joy of Working
Leaders: The Strategies of Taking Charge
Listen and Be Listened to
Managerial Skills for New and Prospective Managers
Managing by Objectives
Managing Conflict
Managing for Results
Managing in Turbulent Times
Megatrends

Products for the Ninety-Minute Hour

No Nonsense Delegation
The One-Minute Manager
Personal Achievement
Personal Selling Skills
The Psychology of Winning
Shirt Sleeves Management
Starting Small, Investing Smart
Strategic Planning
Successful Delegation
Super Stocks
Total Time Management

That's but a sampling of nonfiction, get-ahead, succeed-in-life audiocassettes—all created to help you live a more productive and successful life, and all asking that you devote no time to them while absorbing the information they so conveniently provide for you.

SUBLIMINAL TAPE TOPICS

Although you might be amazed at the completeness of the information now available to you by subliminal methods, I predict you'll end up being equally amazed at their efficacy:

Aging: Paths to Young Thinking
Anger
Attain Good Health
Concentration
Conquer Impotence
Control Spending Habits
Control Stress
Creative Writing
Crystallize Creativity
Curb Excessive Drinking
Curb TV Snacking
Depression
Drug Abuse
Effective Salesmanship
Effective Speaking & Writing
Energy
Enhance Sensory Perception
Fear
Fear of Flying

Fear of Rejection
Finding an Ideal Mate
Finding Your Niche in Life
Imagination
Improve Athletic Abilities
Insomnia
Interpersonal Relationships
Jealousy
Let's Get Organized
Memory Power
Mind and Body Tune-up
Mind Over Pain
Money Attraction
Musical Abilities
No More Guilt
Procrastination
Programmed Weight Gain
Relaxation
Self-actualization
Self-confidence/Self-esteem
Self-healing
Sex Attraction/Sex Appeal
Sexual Satisfaction
Success
Uncondition Smoking Habits
Vivid Dreams
Weight Loss Control

Those are just the nonbusiness titles. Business titles include

Anti-shoplifting
Business Excellence
Company Unity
Developing Managerial Skills
Effective Decision Making
Gaining Executive Status
Money Management

Interested in sports? Then you might want to check out

Baseball
Body Building
Bowling

Golf
Let's Exercise
Running/Walking
Skiing
Tennis

Perhaps you enjoy games more than sports. Then try

Bingo
Blackjack
Craps
Keno
Lotteries
Poker
Roulette
Slot Machines
Sweepstakes Contests
Winning at the Track

Few topics have been overlooked.

Aerobic Fitness
Attract Love Now
Be Decisive
Effective Studying
Enthusiasm
Faster Reading
Financial Planning
Humor
Increase Sales
Joyous Sex (Male or Female)
Letting Go
Look Younger!
Lower Blood Pressure
Money/Prosperity
Passing Exams
Positive Attitude
Positive Parenting
Problem Solving
Real Estate Sales
Safe Driving
Self-control Sugar
Sports Performance

Stop Bad Dreams
Success Network Marketing
Success Strategies
Successful Writing
Super Learning
Time Management
Upward Mobility
Winning Self-image
Worry No More
Yes I Can!

Some subliminal tapes have been expanded into full-scale courses. Although they work well, they do require your time for visualizing, affirming, and experiencing guided relaxation. But in return for that time, you can learn in the following areas:

Accelerating Spiritual Growth
Agoraphobia
Alert and Stress-free Driving
Arthritis
Assertiveness
Attracting Harmony
Awakening Psychic Powers
Becoming a Super Listener
Becoming Efficient and Organized
Becoming Patient
Becoming the Total Person
Body Beautiful
Breakthrough to Success for Students
Breast Development
Bring Laughter into Your Life
Build a Winning Self-image
Cancer and the Whole Person
Confidently Changing Careers
Conquering Bedwetting
Conquering Obstacles
Control Spending Habits
Coping with Emotions
Dealing with Failure
Defeating Discouragement
Dental Anxiety
Develop Foresight and Perspective

Do It Now!
Easier Childbirth
Enjoying Singing
Falling Out of Love
Freedom from Alcohol and Drugs
Headache Relief
Healing and Rejuvenation
I Like to Study
Improving Eyesight
Increasing Good Luck
Intimate Encounters
Intuitive Decision Making
Joy of Pregnancy and Childbirth
Learning to Say No
Letting Go of the Past
Manage Your Time
Overcoming Frigidity/Impotence & Increasing Sexual Respon-
  siveness
Overcoming Shyness
Pain Reduction
Pathways to Mastership
Peak Learning
Positive Expectancy for Realtors
Positive Thoughts for Children
Premenstrual Syndrome
Preparation for Surgery
Psychic Self-defense
Quick Thinking
Radiating Charisma and Warmth
Recapturing Youth, Vitality and Health
Recover Quickly
Sales Achievement
Setting and Achieving Goals
Sleep Like a Baby
Sleep Reduction
Speeding Recovery After Injury or Operation
Start the Day Great
Stay Awake
Stop Eating Junk Food
Stress Management
Stress Reduction
Success for Students

Successful Marriage
Taking Positive Chances and Risks
Talk with the Animals
Teenaging
Trance Channeling
Willpower
Winning at Losing
Working Through Grief

That is but a mere sampling of the available subliminal tapes. Many of the titles listed come in several versions, but I kept it as simple as I could.

Oh yes, if you'd like to hear what subliminal background music sounds like, call 616-538-8338 any time. You'll get a sampling of the musical selections from just one subliminal tape producer. That music is one of the sounds of the ninety-minute hour. Enjoy the music and the extra time.

You can begin to put ninety-minute hours into your life here and now.

If you liked the book, you'll *love* the catalog. For your *free* catalog, call, write, or fax for *The Catalog of Time-Savers,* provided to you at no cost by The Ninety-Minute Hour Institute. It contains the newest and best of the products and services that can fill your life with the maximum amount of ninety-minute hours.

The Ninety-Minute Hour Institute
P.O. Box 1336
260 Cascade Drive
Mill Valley, CA 94942
Call toll-free: 800-748-6444
In California, call: 415-381-8361
Or fax: 415-453-0899